MARINE ELECTRONICS

The Straightshooter's Guide to MARINE ELECTRONICS

Gordon West & Freeman Pittman

With illustrations by Jim Sollers

IM

INTERNATIONAL MARINE
PUBLISHING COMPANY
Camden, Maine 04843

Typeset by The Key Word, Inc., Belchertown, Massachusetts
Printed and bound by BookCrafters, Chelsea, Michigan

Published by International Marine Publishing Company
21 Elm Street, Camden, Maine 04843
(207) 236-4342

Library of Congress Cataloging in Publication Data

West, Gordon.
The straightshooter's guide to marine electronics.

1. Boats and boating—Electronic equipment.
I. Pittman, Freeman. II. Title.
VM325.W46 1985 623.89'3 85-2514
ISBN 0-87742-202-8

Contents

Preface

If we have been successful, what follows is an objective and to-the-point survey of the marine electronic instruments and systems commonly used by sail and powerboat owners and by sport and commercial fishermen. Our goal is to give you the information you need to judge what is out there and what is compatible with your boat and your purposes. Along the way, we will explain the operation, capabilities, and limitations of each instrument.

We'll discuss the ins and outs of installing marine electronics in your boat, and we'll examine a related consideration—whether to buy from a local marine electronics dealer (who will probably install the equipment as part of a sales package) or a mail-order discount house (who will not). We'll also include some procedures for maintaining the gear and for troubleshooting malfunctions. Too many mariners are frightened into inaction by the complex circuitry contained in most instrument housings. Better than 50 percent of all marine electronic malfunctions could be fixed by anyone with a little patience and common sense, and by maintaining your own equipment and doing the simpler repairs yourself (as well as the installation, tuning, and calibrations, when possible), you save money, avoid the inconvenience of equipment "down time," and become a more skilled and knowledgeable operator.

Limitations of space dictated against the inclusion of a few marine electronic items we would have liked to discuss. Weather facsimile receivers, for example, are proving invaluable for commercial fishermen and for offshore racers and cruisers, but because they have been installed in fewer than 4,000 vessels to date, we did not include them here. If, as we hope, this guide should become an established reference, updated every one to two years, then we will perhaps include them in the next edition. Similarly, we do not discuss the Omega system of radio navigation, which has taken a back seat to loran and satellite navigation, and there are corrosion control systems, shipboard security systems, and other miscellaneous members of the marine electronics family that we might have included, but didn't. On the other hand, we did make room in the single sideband radiotelephone chapter for a discussion of ham radio, in the hope that this book will clear up once and for all the confusion regarding the proper uses of VHF, single sideband, and ham transceivers. By the same token, the sailboat performance instrument chapter includes integrated systems and add-on computers.

This guide does not include model comparisons. To do so would have required independent testing facilities, and in any case, the marine electronics marketplace is evolving so rapidly that at least some such comparisons would have been out of date by the time the book was published. Instead, we have attempted to provide all the information you will need to compare and evaluate models yourself in an informed fashion. This knowledge should be as current in 1986 as it is in 1985.

Perhaps the most important consumer information in this book concerns the question of where to buy a marine electronic item or package. The most frequent choice is either a local marine electronics specialty dealer or a mail-order discount establishment. The former are too numerous to list; you can probably find a good one in your locale. Chapter One gives you some criteria to go by, but don't forget the importance of word of mouth. The mail-order discount houses are listed at the back of the book, along with the manufacturers of the equipment we discuss. You can write the latter directly for information on gear that interests you.

Never before have we had so much good marine electronic equipment to choose from. This equipment is much smarter, more compact, and in many instances *less* expensive now than it was just a few years ago. Not surprisingly, the process of shopping for gear is correspondingly more complicated. We hope this guide helps.

1 The Marine Electronics Marketplace

by Gordon West

Probably the most frequently asked question in the marine electronics business is, "Which unit is best?" It's a good question, but there is no one right answer. The question should perhaps be rephrased as, "Which marine electronics are best for my particular application?" To answer this properly, the marine electronics seller would need to ask you many questions about your boat, your intended cruising or fishing area, your experience in operating electronic equipment, and the level of sophistication you desire. If any seller tells you which unit is best without interrogating you first, look for another place to shop.

DECIDING WHERE TO BUY

There are four major avenues of distribution for marine electronic equipment:

1) The traditional marine electronics dealer
2) Marine electronics catalog houses
3) Marine hardware stores that also sell electronics
4) Boat dealers who include built-in electronics with new or used boat sales

There is no love lost between the specialty dealers and the catalog houses. The sales and servicing of marine electronics have traditionally gone exclusively to the former, but the catalog houses have recently made major inroads on the dealers' sales.

Most specialty dealers have elaborate showrooms with examples of every type of marine electronic gear hooked up and operating. This allows you to hear, manipulate, and get the feel of any type of equipment you may be thinking of purchasing, and there is no substitute for this. Most dealer showrooms also stock an ample supply of literature for each piece of equipment that is on display, as well as some that are not.

The specialty dealers hire the cream of the crop when it comes to qualified sales personnel. Many of the men and women who make up their sales forces are qualified not only to recommend gear but to give you a fairly technical overview on any piece of equipment in their showroom. In addition to these sales personnel, a dealer usually employs three to eight FCC-licensed technicians who possess marine electronics technician certificates; these are the top experts in their field, and many of them have had extensive additional factory training on specific products. (Equipment manufacturers spend thousands of dollars training technicians on the installation and operating techniques and the servicing of particular pieces of equipment.) If you have a question that the dealer's sales force can't answer, they can often bring out the technician "from the back" who will answer your query in detail. The dealer can also provide you with a custom installation aboard your boat, or assist you with a do-it-yourself installation.

The marine electronics specialty dealers also offer factory-authorized warranty repair. Thus, if your new piece of gear should fail, it may be fixed for free by the dealer. Sometimes the equipment must be taken back to the shop for service, but in other cases the fix can be accomplished aboard the boat. Troubleshooting of equipment aboard your boat is another unique feature of a specialty dealer. They have the portable equipment to do it. Labor rates are high if not covered by warranty, but are usually worth every penny. (More about warranties shortly.)

Specialty dealers have united with manufacturers to form a trade organization called the National Marine Electronics Association (NMEA). The NMEA has a code of ethics that all member dealers must meet; if a dealer sports the NMEA logo, you can be sure that he is indeed a specialty dealer who has met the membership requirements.

Marine dealers normally sell their equipment at the manufacturers' suggested retail prices. They are not known for heavy discounting, for they must maintain a minimum markup in order to cover the showroom, the trained sales personnel, and the expensive service equipment. It is for this reason that catalog sales of marine electronic equipment have been growing so strongly in recent years.

Marine catalog houses that sport exciting color brochures and even more exciting discount prices are indeed doing well. These relative newcomers to the world of marine electronics got their start when gear that could be user-installed became more common, and when the manufacturing of marine electronic gear began to overwhelm the available avenues of distribution. Twenty years ago most marine electronic equipment was American-made, and production was just right to support the specialty dealers. There was no such thing as too much gear to sell, and discounting was taboo. Now United States manufacturers of marine electronics are importing a huge portion of their gear from the Far East. In doing so, however, the domestic manufacturers have lost control of production volume. The Oriental manufacturers don't and won't slow down on production, so alternate sources of fast distribution are necessary to keep the pipeline clear. The name of the game for the marine electronics catalog-discount house is quick sales, which they achieve with a low price. Some very aggressive promotional work is carried on by the marine electronics discount catalogs and showrooms. Their creative marketing skills leave the specialty dealers in the dust. Manufacturers must surely be impressed with their sales figures, as well as their promotional ideas.

If you plan to install your equipment yourself, and you know what you want, discount houses and catalogs might be a good way to go. But don't expect many catalog houses to give you an hour or two of their time going over a particular piece of marine gear. Their toll-free numbers are for placing orders, and their personnel, though they may indeed be knowledgeable about marine electronics, are not on the phone to plan your entire marine electronics system. Be advised, too, that discount and catalog houses frequently don't tune or repair the equipment they sell, and most don't offer any installation service at all. Advice, yes; service, no. Several marine electronics manufacturers will not include the necessary FCC certification with their radiotelephones or radar, and a factory-authorized and licensed dealer must either install the equipment for you or inspect your installation prior to "signing off" on the FCC radar or radiotelephone certificate. If you buy the equipment from a discount house, this will be an additional out-of-pocket expense. If your equipment breaks, most catalog houses will tell you to ship it back to the manufacturer for repair. This "we sell it, they fix it" approach may not be popular with manufacturers, but when a lot of units are sold through these houses, they don't have much choice.

Purchasing a piece of marine electronics through a discount house, having it fail, and then taking it to the marine electronics specialty dealer down the street for repair may cause hard feelings, especially if the specialty dealer took the time originally to tell you which piece of equipment to buy. Some dealers might throw you out the door. Others, if they have a repair shop, may fix your unit, but most will take their sweet time if you bought it elsewhere.

A few marine electronics discount organizations do possess service as well as installation facilities. These are rare, however, and you should carefully check out such an organization before buying equipment from them. You may wish to use the checklist at the end of this chapter for that purpose. Even if a discount house staffs two or three technicians, these are probably unable to keep up with the repair work generated by the discounter's large volume of sales. As a general rule, when you buy equipment from a catalog house, you will encounter less delay and frustration if you send it directly back to the manufacturer for repair.

Buying electronics in a marine hardware store is downright questionable. What you will find will be a gaggle of equipment sealed under the glass in the showroom counter. Rarely is it hooked up, and seldom will the hardware store personnel know anything about it. If the equipment should fail, they won't be able to service it. They will usually send you down the street, where you should have bought it in the first place. Marine electronic gear is getting as sophisticated as a personal computer. Let the hardware stores sell their cleats, lines, and boathooks, and don't buy marine electronics through any store that can't tell you how to install, run, or fix them. There may be some exceptions, but most hardware stores have no expertise in this area.

Buying marine electronic equipment with a boat and having it installed by the boat dealer is another proposition. Most boat dealers will subcontract the installation to local marine electronics specialists, so you can get a good deal in a packaged boat sale. Just be sure to find out who's going to do the installation of the equipment, who will service it, and who in the world is going to show you how to run it. Most boat dealers are specialists in boats, not electronics.

If you are buying a boat that already has marine electronics aboard, make a list of the gear and take it to a local marine electronics specialist to find out how up to date it is. The better marine electronics specialists will give you a fair answer, but you may want a second opinion. New boats with built-in marine electronics may be a very good deal if the electronics are exactly what you are looking for. The more serious mariner or marine electronics enthusiast, however, will specify certain pieces of equipment. If such a request can be honored, make sure a qualified technician will do the installation.

The retail margin from the manufacturer to the seller of marine electronics averages about 40 percent. On large equipment such as radars, water makers, and sonars, it is usually less, and on smaller pieces of equipment such as VHF transceivers, depthsounders, and RDFs, the margin can be as much as 50 percent. Marine antennas may go as high as 60 percent. Keep this in mind when you are offered a percentage discount off the manufacturer's suggested retail price.

The marine electronics specialty dealers like to operate at a 30 to 40 percent margin, while the discounters can operate as low as 10 to 15 percent above cost if they move large quantities. Boat dealers that offer marine electronics as a package may sometimes break even—the marine electronics will help them sell the boat, and that's their main interest. This gear will not be too heavily discounted, however, since the boat dealers buy from another distributor rather than directly from the manufacturer. The hardware stores, too, usually buy from distributors, the distributor making about 10 percent and the hardware store hoping to sell the gear at a margin of 30 percent.

Summing up, if you want a turnkey operation and the ability to put your finger on one person to select the appropriate equipment, install it, and show you how to use it, the marine electronics specialist is a good way to go. If you want the best buy for the least money and you plan to install and tune everything yourself in your spare time, you might want to buy your gear through a discount house or catalog.

If it's your wife's birthday and a Sunday to boot, and she absolutely has to have that marine VHF handheld radio, and you spot it in an open marine hardware store, then go ahead—pay the retail price, get your set, and hope it doesn't break.

If the boat price is no longer negotiable but you can get the dealer to throw in a bucketful of marine electronics, have at it.

Equipment installation is an important factor. Unless you're a confirmed do-it-yourselfer or an electronics hobbyist, for example, you're probably going to want to have your radar system, single sideband station, or automatic direction finding station professionally installed aboard your boat. If you buy the equipment from a mail-order house and then ask a local marine electronics dealer to do the installation, you'll be charged on a time-and-materials basis. Using SSB radio as an example, a typical installation takes two days or more, and at $40 per hour plus materials you would be facing an additional out-of-pocket expense of at least $700. If, after the unit was installed and certified, you were not satisfied with its operation, the additional grounding or other recommended remedy would cost you more money still. If, on the other hand, you buy the SSB radio (or radar, or ADF station, or loran, or . . .) from the local dealer, you can have the price of the installation included as part of a sales-and-service package. Have this agreement drawn up beforehand, in writing, and insist that it permit you to call back the technician *without additional charge* if you are not happy with the initial installation. An installation included as part of a sales agreement should cost substantially less than one contracted on a separate, time-and-materials basis.

Probably the best approach is to buy larger, hard-to-install, hard-to-remove gear from the marine electronics specialist and smaller, removable equipment from a discount house. If the big stuff breaks you can bring the specialist aboard to fix it. If the little stuff goes sour, you can unplug it and either take it into the discounter's local repair depot or ship it back to the factory. The discussions of individual instruments in this book will give you a feel for which you may safely buy from a discounter.

The most important thing to remember, no matter where you buy your equipment, is not to trust anyone who says, "This equipment is best for you," without first finding out how you plan to use it. Take time to look at your options when buying marine electronics, and always play fair—don't pick the brain of the marine electronics specialist only to turn around and buy your gear through a discount house. Stay honest with the people you are dealing with, and you may find that they will meet you halfway when you tell them what you want to put aboard and what your budget is.

One further warning: No matter who you buy from, you are definitely taking a chance when the new equipment you buy is delivered in a carton that has been previously opened without any documentation having been added. Some sellers check out and certify equipment before selling it, and if this is the case, that's fine. But demand their certification sheets on the equipment in the preopened box. Otherwise, you may be getting a unit that has already been sold once. It is not unusual for a buyer to return a unit to a dealer complaining of one defect or another. If the dealer can't find the defect on the test bench, he may decide the buyer was mistaken, repack the unit, and sell it to the next customer, who might be you.

WARRANTIES

Warranty periods for free repair and possible retuning of the equipment you buy may be as short as 90 days or as long as two years. Study the warranty before you buy, because there are plenty of loopholes.

The manufacturer's warranty is expressed clearly enough in the paperwork that comes with each piece of equipment. If you take the unit off your boat and send it back to the manufacturer (who is in reality the American importer if the unit is manufactured in the Orient), they will fix it free of charge provided the warranty has not expired. You pay the shipping charges, however. Most factories will get the unit back to you within three weeks. But if it's not convenient or even feasible to return a piece of gear to the manufacturer for repair, you will be dependent on a factory-authorized, licensed repair facility, and here you may encounter delays or frustration.

As discussed, not all sellers of marine electronic equipment are licensed repair facilities. If you buy from a discounter, for example, you may have to go somewhere else to have the unit repaired or tuned. If that somewhere else is a specialty dealer, he may or may not do the repair, and he will almost certainly put your gear at the end of the line behind any equipment bought directly from him. He will probably refuse to honor the factory-authorized warranty if you do not have proper sales slip information, so do send in the warranty cards and do keep accurate purchase records.

Marine electronics specialty dealers and a few discount catalog sellers will service everything they sell, and are factory-authorized to do so. They will give you a written statement on the terms of their free service, and they may have extended-service options as well. These written statements should be examined carefully. While manufacturers will often pay dealers to service your equipment when it is under warranty (many demand that the more complex units, such as SatNav receivers, be returned to them for repair), they may not pay a technician's travel expenses to your boat. Will a dealer pay for that service if you buy the unit from him? Find out beforehand; a permanently installed radar unit would not be convenient to dismantle and return for service.

Remember, too, that even if a discount catalog seller is factory-authorized to service and repair the equipment, you will usually pay the shipping charges, and you would probably get your equipment back sooner were you to ship it directly to the manufacturer.

When you buy equipment, ask what will happen should your gear fail in a distant port. Can the seller or manufacturer supply you with a list of authorized service facilities that will honor your warranty? For reasons already stated, this is particularly important when you buy mail-order equipment, but it applies when you buy from a specialty dealer, too. Although specialized marine electronics dealers are united through the NMEA, a dealer 1,000 miles away from where you bought the unit will not necessarily honor the warranty and fix your set for free.

Get your warranty claims in writing, and more important, find out exactly what the local marine electronics dealer will do to back up the manufacturer's warranty. Some enterprising dealers will exchange any item that fails within the first 10 days of operation. That's a nice feature to look for.

COMPARING MODELS

There are no specific model comparisons in this book, because the marine electronics marketplace changes so rapidly that individual comparisons would be out of date by the time they appeared in print. In December 1984 there was only one fully synthesized SSB radiotelephone available for less than $2,000, while two months later there were three. This is but one of many similar examples.

Most of the changes in a given type of marine electronic gear are small and incremental, however, and the fundamental criteria by which the gear should properly be judged change more slowly. It is on that premise that the consumer advice in this book is based. For each marine electronic instrument or system, the criteria of selection are set forth. Follow these criteria carefully, supplementing the process with input from dealers and fellow mariners, and the gear you buy will be appropriate for your purposes.

The National Marine Electronics Association gives annual awards to the "best piece of marine electronics" in each category, these awards being decided by the dealers. But although the quality and capabilities of the equipment are weighed in the selection process, the dealers are biased toward units that give them a good profit margin, and the awards are thus, in part, a popularity contest. Also, marine electronics dealers have been slow to accept new and innovative designs, preferring to wait as long as several years while the designs prove themselves. Ten years ago, when the first keyboard-entry VHF marine radio was introduced, it represented a breakthrough, but it won no industry awards or accolades partly because the marine electronics specialists simply didn't trust a radio without knobs, and partly because they wanted a more favorable profit margin on the product. It is true that new designs sometimes need a year or two to stabilize and have their bugs worked out (and you can use this as a criterion for selection if you don't mind buying last year's technology), but there are nevertheless new instruments each year that deserve awards but garner none. The NMEA awards signify marketability first and foremost, and true excellence only incidentally.

The quality of the instruction manual is a good clue to the quality of the marine electronics unit you may be planning to buy. Better units will have a more complete instruction book, and some units will have an additional service manual that may be purchased separately. If you can't quite decide which piece of electronics to choose, ask the dealer for a peek at the instruction book.

A technical service manual is a boon if you plan to have the equipment serviced by different dealers along the coast. This eliminates the age-old problem of dealers telling you they can't fix your equipment because they don't have the appropriate service manual. The service manual might also give you some basic troubleshooting guides to follow should your equipment go on the blink. Again, the more complete the service manual, generally the better the equipment.

Most important, however, the best electronics are those which fit with greatest precision your installation and operating needs. Again, if a seller recommends a particular model before asking you any questions about how you plan to install and use it, he is only selling equipment, and not a service. Don't decide on the name brand until you have seen them all and reviewed the capabilities of each.

BUYER'S CHECKLIST

Is the marine electronics seller you are considering characterized by the following?

- Factory-authorized service approval
- Sells marine electronics as its major business
- Buys directly from the manufacturer
- Offers trained salesmen to spend up to one hour with you in planning your system
- Has demonstration equipment on the floor
- Offers catalog sheets for you to study at home

- Has a licensed technician available for consultation
- Offers installation and service aboard your boat
- Possesses the necessary technical equipment and personnel to calibrate, tune, test, and FCC-certify all equipment
- Offers a selection of equipment
- Stocks the popular accessories for each unit
- Boasts factory achievement certificates of technical merit for technicians

Here are some other questions to ask:

- Is the equipment sold to you in factory-sealed containers, or is it "demo gear" taken home by someone else?
- Can the seller modify the equipment to meet particular installation needs?
- Will the seller replace dead-out-of-box equipment?
- Will the seller loan you a piece of equipment while your new piece is getting fixed?
- Can the seller provide a custom installation, such as cabinetry, for your equipment?
- Is the seller a member of the NMEA?
- Is the showroom or catalog tidy, complete, and professional?
- Has the seller been in business for more than a few years?
- Can your purchased equipment be fixed in any other cruising area in the United States or the world?
- Is the pricing competitive?

If you can answer "yes" to most of these questions, you have found a good, safe place to buy. For small pieces of equipment, if you know what you want, you can accept "no" answers to more of these questions as you gravitate toward reputable discount houses in search of a better price. If you *do* buy from a dealer, demand the strong warranty backup for which you're paying, and demand that the Loran-C receiver, SSB radio, SatNav receiver, radar set, autopilot, or automatic direction finder be properly tuned, calibrated, tested, installed (if dealer installation is part of the package), and, if necessary, FCC-certified.

2 Depthsounders

by Gordon West

There have been huge advances in depthsounder technology within the last five years. Foreign and American depthsounder manufacturers are battling it out to see who can come up with the most features at the best price, and we consumers reap the benefits. The latest generation of depthsounders will work at higher boat speeds, suffer less ignition interference, ignore swiftly passing air bubbles, and stay locked onto the bottom for up to five seconds when your boat pops completely out of the water at 50 knots. Yet prices are lower than they have ever been. Two distinct types of depthsounder have emerged, one for navigation and one for locating fish. This chapter covers both.

HOW THEY WORK

The principle of operation behind echo sounding hasn't changed over the last 50 years. A depthsounder beams pulsed sound waves down toward the bottom, then picks up the weak echoes and converts this information into bottom and fish readout information, which is displayed on the indicator. While variations in water temperature, salinity, and density affect the speed of sound somewhat, acoustical pulses normally travel at about 4,945 feet per second in water, compared to 1,117 feet in air. If it takes two seconds for a pulse to leave the underwater sender/receiver unit, bounce off the bottom, and return to the unit, the depth of water must be 4,945 feet.

Depthsounders consist of three components:

(1) The *transducer.* This is the underwater device that translates pulses of electrical energy into acoustical vibrations, directs these toward the bottom, receives the echoes, and converts them back into electrical impulses. The transducer case may be all plastic or bronze with a plastic face, and it is mounted with its face in the water and pointed downward.

(2) The *amplifier.* Located inside the depth indicator unit, this component actually comprises three subcomponents: an oscillator to produce electrical impulses of a uniform frequency; a transmitter amplifier to step up the impulses before sending them to the transducer; and a receiver amplifier to pick up the faint return echoes and convert them into strong, usable signals that are fed into the display or indicator circuitry. Bigger and more expensive transmitter amplifiers give stronger signals to reach extremely deep bottoms.

(3) The *indicator.* Take your pick! Chart recorders,

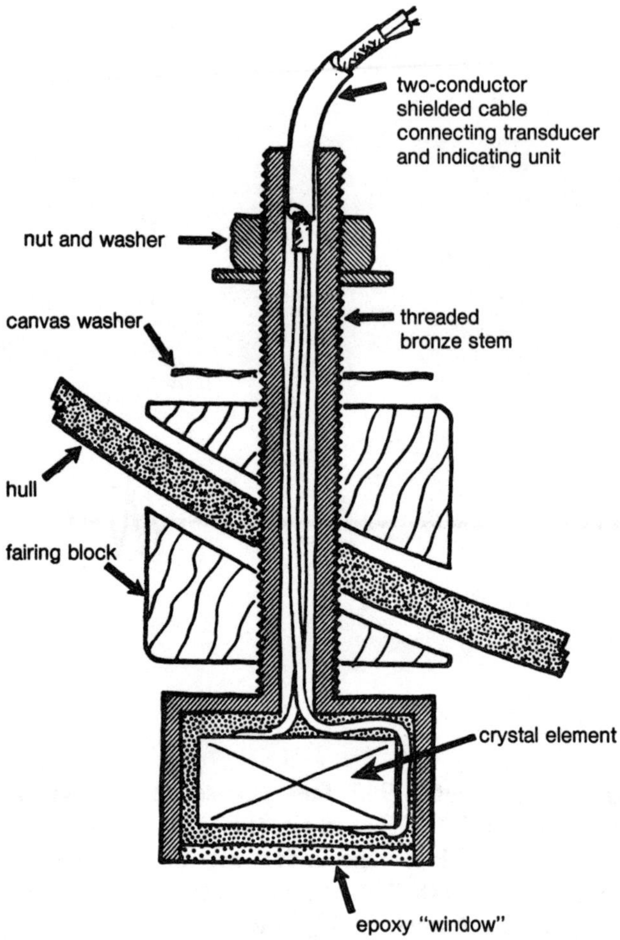

Cross section of a through-the-hull transducer mounting, showing the fairing blocks that must be hand-carved to fit the hull shape precisely, while keeping the transducer in a vertical position.

flashing neon bulbs, liquid crystal displays, video presentations, and other readouts are available for the indicator unit. This is the device that translates the transducer input into a readout form—graph, digital, or analog—that you can understand and interpret. When you select one of the several types of sounders, what you are really choosing among is the various types of indicator units.

The transducer is a critical part of your depth-sounder assembly. Many different types are available with any one type of sounder, and the one you choose must fit your particular echo sounding needs. It contains a precision firm slice of quartz crystal buried beneath a plastic membrane. When a voltage pulse is applied to the crystal, it will vibrate at its designed frequency. The vibration is called a piezoelectric oscillation, and it can be made to recur many hundreds of times per minute through the application of alternating voltage between two metal plates that sandwich the crystal. In the majority of depthsounders the quartz oscillates at a frequency of 200 kHz, or 200,000 cycles per second, well above the range of human hearing. Depthsounders that need to send a signal extremely deep will use a lower frequency crystal, usually at 50 kHz (50,000 cycles).

These sonic waves are aimed downward, reflect off the bottom or from fish, and then reenter the underwater transducer assembly, which picks up the weak return signals between transmitted pulses. The incoming echo strikes the crystal (also called a "plate" or "element") and applies a tiny mechanical stress that travels to its opposite face; the crystal is set vibrating mechanically for an instant, and in doing so generates an alternating voltage between its two faces. The vibrating sound pulses are thus turned back into electrical energy that is then fed through the metal plates and back to the receiver amplifier. Shallow water sounders may generate up to 25 watts of peak power (from the transmitter amplifier) so that the sound pulse will strike the bottom with enough force to return a substantial echo. In deepwater units, over 100 watts of power must be used to pick up the distant bottom, this power applying more than 400 peak volts to the transducer's piezoelectric mechanism.

The pulse repetition rate will vary between and within echo sounders depending on how deep they are intended to sound. On shallow sounding units, pulses are transmitted and received up to 1,500 times per minute. On deeper sounding units, more time is needed for the echo to be returned, so the pulse repetition rate may be as low as 60 times per minute. This lets the receiver section listen for the weak echo before the transmitter section fires off another ping. Shallow sounders will also transmit short pulses, and deeper sounders will transmit longer pulse lengths. For an analogy, imagine yourself in a boat in the fog, giving blasts on a foghorn and listening for echoes from a distant shore. Because the sound may be scattered by the shore and the fog, and intermittent background noises may intrude, you'll stand a better chance of picking up the echo of a longer blast. On most sounders offering more than one depth range to select from, pulse length and repetition rate are both set automatically when you "dial" the depth.

A 200 kHz transducer can be used for fishing, navigating, or both. Like transducers of other frequencies, it sends out a cone of acoustical energy that gets wider as it goes deeper, the usual beam width at 200 kHz being 9 to 15 degrees. (Beam widths up to 45 degrees are available for shallow water fishing.) With a 9-degree beam, a 200 kHz transducer is effective at depths up to 600 feet; at 100 feet, it will "see" a circle about 14 feet in diameter, giving quite a detailed picture of the ocean floor. If fish are right under you—even individual fish of moderate size—the 200 kHz, 9-degree transducer will pick them up quite nicely. Most digital indicators (which are used strictly for piloting) and flashers (which have some fishfinding capability) use 100 to 200 kHz transducers.

For a larger picture of the ocean floor, you need a physically larger, lower frequency transducer. The 50 kHz transducer may have a beam width of up to 54 degrees, covering a 100-foot-diameter circle in 100 feet of water. This lets you see fish activity off either side of your boat, and the lower frequency enables you to sound much deeper bottoms (3,000 feet or more, depending on the transmitter power). You lose a bit of definition, however; you won't be able to pick out those small rocky protrusions on the ocean floor, as you can with a tighter beam angle, and small fish may escape detection. For somewhat improved resolution at depths beyond the reach of a 200 kHz sounder, a 50 kHz transducer with a narrower beam width (as narrow as 16 degrees) might be a better choice. Regardless of beam angle, however, no 50 kHz transducer will provide the shallow water resolution of a 200 kHz transducer. The 50 kHz transducer is commonly used where both navigation and deepwater fishfinding capabilities are desired.

If you want the best of both worlds, choose a depthsounder with two transducers—perhaps a 200 kHz transducer with a wide cone angle for shallow water fishing, and a 50 kHz transducer with a narrow cone angle for deepwater fishing and sounding. You can use a transducer switch box to select between them. Fishfinders frequently use this configuration.

Check with the local fishing community to see what type of transducer they are using for a particular fishing area. If you already own a depthsounder, you can switch transducers providing you match the new transducer frequency and impedance with the frequency and power of your depthsounder transmitter. Check your instruction manual for the operating frequency of your sounder. The converse is also true—you can replace your indicator unit with a "smarter" or more sophisticated unit (from a different manufacturer if you desire), yet keep the old transducer. Again, you must ensure that the frequency and power of the depthsounder transmitter matches the transducer. If the transducer is physically small in size, it could be damaged by an extremely high-powered transmitter. High-power, low frequency transmitters, such as those used for extremely deep sounding, require massive transducer systems and will not be compatible with the typical 200 kHz transducer. With this exception, however, mixing and matching components is quite feasible and certainly advantageous, since transducers seldom wear out. You may not get a discount on the indicator unit by telling the seller you don't want the new transducer, but at least you'll have a spare transducer, and you'll save the hassle and expense of another transducer installation.

TRANSDUCER INSTALLATION

The transducer must be in contact with the water for best performance, and proper location and fairing of the installation is critical. On larger vessels, power or sail, the transducer is usually mounted through the hull. Although manufacturers' instruction manuals for other types of marine electronic equipment usually offer complete installation directions, this is often not true of depthsounder transducers mounted through the hull, because the manufacturers assume, wisely, that you will leave the job to a boatyard. If you do decide to go it alone, however, the key to good depthsounder performance is to find a spot where no aeration will disturb bottom soundings.

The placement in a sailboat must be far enough from the keel that signals are not deflected when the boat is heeled. To get around this you can mount a transducer either side of the keel with a gravity switch automatically selecting between them. Many available depthsounders can be modified for this system.

The transducer is mounted through the hull using a special fairing block that will keep it pointed directly down toward the ocean floor when the boat is on an even keel. These are streamlined to minimize drag and to shield the transducer from turbulence, bubbles, and damage. Sounders that are professionally placed through the hull usually give superb performance, and

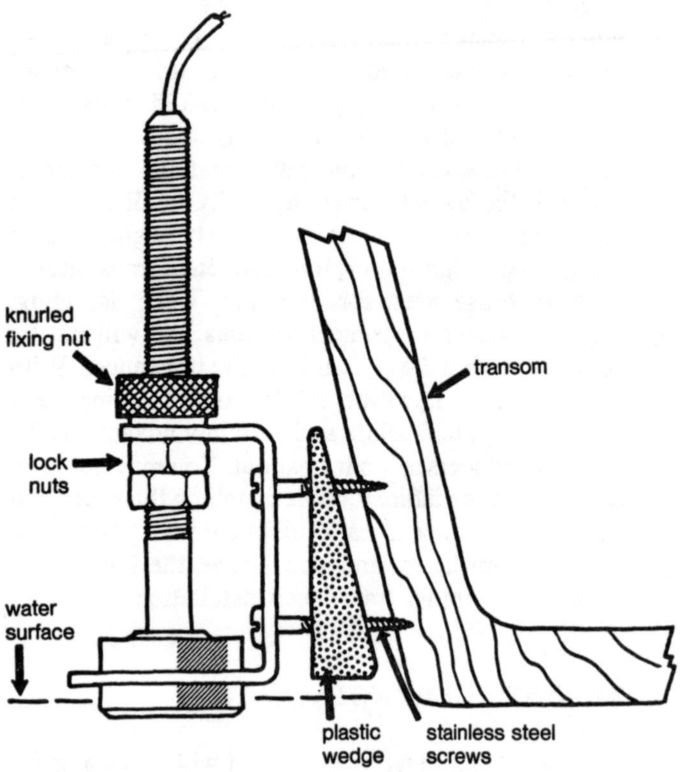

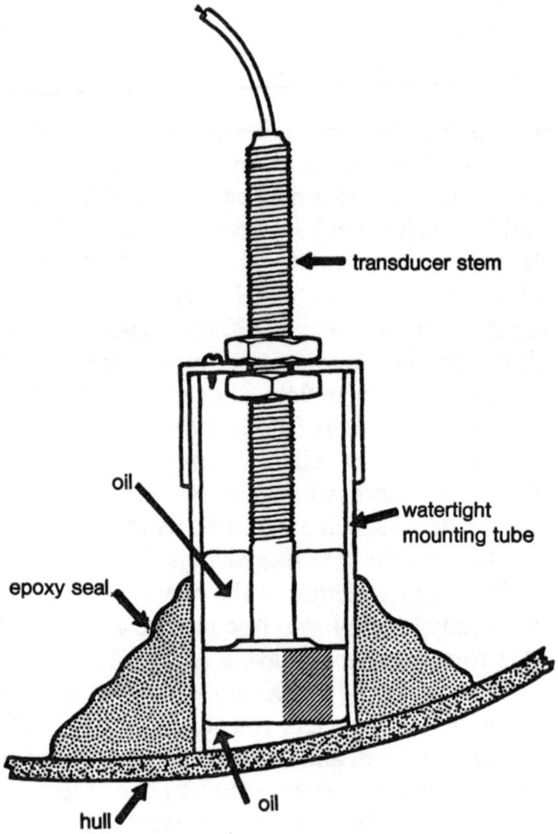

Left: *Transom-mounted transducers are subject to erratic operation caused by air bubbles flowing off the boat's bottom. They are practical for slow speed boats where the transducer is totally submerged at all times, but on planing hulls they may provide only intermittent information. Some brackets are gimballed, and by pivoting the transducer it is sometimes possible to find a position that will minimize erratic readings.* ***Right:*** *There are various means of mounting a transducer inside the hull, but almost always, the installation must be custom fitted. The transducer must be positioned vertically and the face of it submerged in oil to exclude any air between it and the hull. Any transducer with a threaded stem can be installed in such a watertight compartment.*

if you are into serious long-distance cruising, this is the only way to go. But let a professional (often the boatyard) do the mounting unless you're sure you know what you're doing. Transducers *can* be mounted by a scuba diver with a minimum amount of water let into the bilges, but if you want to try this trick, get an experienced diver. It would be much better to have the transducer mounted during your next haulout.

Most manufacturers supply a transom bracket for small boats that allows you to mount the transducer off the stern (but still underwater), and some such brackets allow the transducer to self-adjust its orientation depending on your speed through the water. A slight miscalculation of transducer placement can seriously compromise readings even at moderate speeds, however, and on planing hulls the readings may be intermittent.

On some boats, you can mount the transducer inside the hull and sound through the wood or fiberglass. The accuracy of readings will not be affected, but the maximum sounding depth will be reduced by as much as 75 percent or more, and the ability to find fish may be all but eliminated. This system will not work at all with most cored hull constructions or when the laminate has local voids. Nevertheless, it may be the method of choice for racing sailors who are greatly concerned with hull drag, or for high-speed powerboats.

If you decide to try an inside-the-hull mounting of the transducer, follow the manufacturer's instructions precisely. The usual first step is to try out the transducer at different points below the waterline. The face of the transducer must be flat against the wood or fiberglass hull and covered with water or oil. Putting your boat in fairly deep water, move the transducer around in the bilge and find a spot where you get a solid reading of the bottom. If you find the results satisfactory, you can spring for one of the commercially available kits designed specifically to mount any type of transducer to the inside of your hull.

One further note: the marine electronics specialty dealers do many installations, but they will *not* install transducers. This job is usually done at a boatyard.

The transducer assembly is linked to the depthsounder equipment by a two-conductor coaxial-type cable. This cable should never be nicked or cut. Allowing moisture to enter a crack in the outside black sheath could damage the entire affair. Since this cable is carrying both transmitted and received signals, it must be shielded from external noise. *Never* route it beside the primary wires in your ignition system, and while you *can* run the cable beside other electrical wiring for short runs, it's best to keep it as far away from them as possible. Here are some examples of wires you want to avoid:

- Wires to fluorescent lights
- High-voltage wires from your SSB radiotelephone antenna circuit
- Satellite antenna wires
- Bait tank wiring

The transducer cable usually contains an outside braid to keep static out and the signal in. Should the cable be severed, you must carefully reattach all of the wires, making sure that the braid covers up the entire works.

The end of the transducer cable simply plugs into the back of your depthsounder. Watch this connection! If neglected, it may become loose, corroded, and then intermittent. Make sure also that none of the wires pull free from the chrome connector collar. Intermittent bottom readings at slow speeds are sometimes the result of a loose or poorly connected transducer plug.

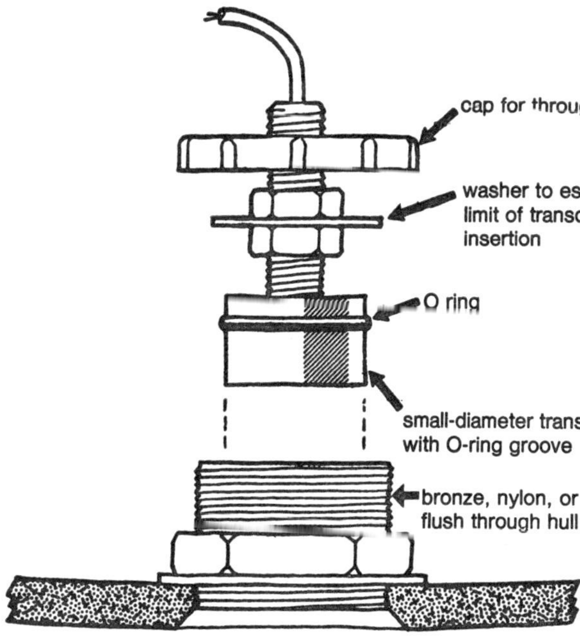

In waters where there is a heavy growth of barnacles, it is sometimes necessary to remove the growth from the face of the transducer to keep it operating properly. Special through-hull fittings are available that permit withdrawal of the transducer into the bilge for cleaning. One must screw a cap over the fitting when the transducer has been removed. A small amount of water will rush into the hull during the moments the transducer is being withdrawn and before the cap is screwed down.

INDICATOR UNITS

Some manufacturers refer to their depthsounders as "sonars," but this is really not accurate. A true sonar is a device that sends out energy horizontal to the surface of the water, while a depthsounder, of course, directs its energy perpendicular to the water surface. Sonars are very expensive pieces of equipment designed for commercial fishing applications—spotting large schools of fish, rocks, or obstructions on either side, astern, and ahead of the vessel. Depthsounders won't do this. Even if you turn a depthsounder's transducer horizontal to the water, it won't do it, because it does not possess the right type of transducer and associated circuitry to scan horizontally without picking up the water surface as a false target.

Manufacturers have developed extremely powerful transmitters and extremely sensitive receivers to be housed in the indicator unit, but don't be overly influenced by advertised transmitter powers. Since these claims may refer to peak watts, peak-to-peak watts (a highly inflated figure), average watts, or useful watts, comparisons are difficult. For better or for worse, the better indication of the performance of a sounder is its price tag—the higher the price tag, the more watts in the transmitter, and generally the greater the sensitivity—that is, the ability to see weak echoes in the receiver.

The difference between a depthsounder and a fishfinder lies in the indicator unit and the micro-

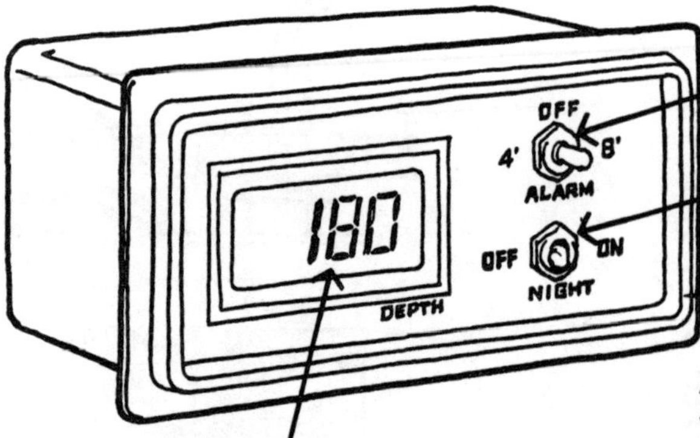

A digital depthsounder. Easy to read, but will not show fish. Digital sounders are usually weatherproof and may be mounted in the cockpit. The price range (including transducer) is $200 to $500, and a typical indicator-unit size is 3 inches high, 7 inches wide, and 4 ½ inches deep. A meter-reading or analog unit (not illustrated in this chapter) is slightly larger.

processor circuitry associated with a particular type of display.

Digital readout indicators are best for anchoring and handy for piloting, and they are the frequent choice of cruising and racing sailors and other consumers whose concern is depth in feet, fathoms, or meters, not where the fish are or what the bottom is like. Manufacturers are getting away from duller, light-emitting diodes (LEDs) to the more brilliant liquid crystal display (LCD) readouts, which can be seen easily in direct sunlight and are backlighted for night viewing. The LCD readout also draws less current, so your depth indicator can operate off portable dry cell batteries. These readouts are reliable and efficient and are now considered "state of the art." Light-emitting diodes are fast disappearing.

Digital sounders and meter-movement (also called "analog") sounders are both easy to read. In addition, digitals have no motors or moving parts, so they operate silently and effectively. Some digitals sport innovative features such as depth alarms and anchor alarms, products of microprocessor technology. You need only predetermine at what depth or depths you wish the alarm to sound. If you aren't watching the unit and get too close inshore, your alarm will let you know immediately. If you're asleep at night and are blown off- or onshore, the alarm will sound off letting you know you have exceeded a minimum or maximum preset depth setting. Sounders offering this feature differ widely in the available range of depth alarm settings. "Forward looking" sounders will even keep track of the rate of change in depth, giving a warning alarm when the water shoals rapidly toward a critical depth. Digital indicators with averaging circuits are advertised, and these will ignore wildly errant, ephemeral readings caused by debris or bubbles, or by a powerboat's hull momentarily lifting from the water at a high speed. Some digital sounders will automatically adjust raw depth readings (which, by definition, indicate depth below the transducer) to depth below the keel or below the water's surface. Some units are wired to speakers and programmed to deliver vocal depth reports.

Flashing indicators or flashers utilize a neon bulb or LED mounted on a rotating disk that is driven by a constant-speed motor. On each rotation a switch is magnetically activated, causing the transmitter to send an electrical pulse to the transducer. When the weak pulse from the returning echo is amplified, the bulb or LED is lit up and displayed against a circular calibrated scale. Because the disk rotates at a constant rate, the arc swept by the bulb in the time between outgoing and returning pulses is directly related to the depth as read off the scale. The disk rotates so rapidly that the flashes appear motionless but for changes in depth. Random flashes usually indicate fish, and if you watch a flasher closely, you can see the return echoes of even one or two large fish just off the bottom. Flashers are sometimes used by sportfishermen because they are inexpensive and they work well providing you keep your eye on the dial. Most flashers give you a number of depth ranges to select from. In making a selection you are in actuality adjusting the speed of the disk's rotation, with slower speeds being required for greater depths.

Flashers will also indicate the type of bottom below you. If you receive a broad bottom flash, it's probably mud. A sharp bottom flash is likely to be hard sand or

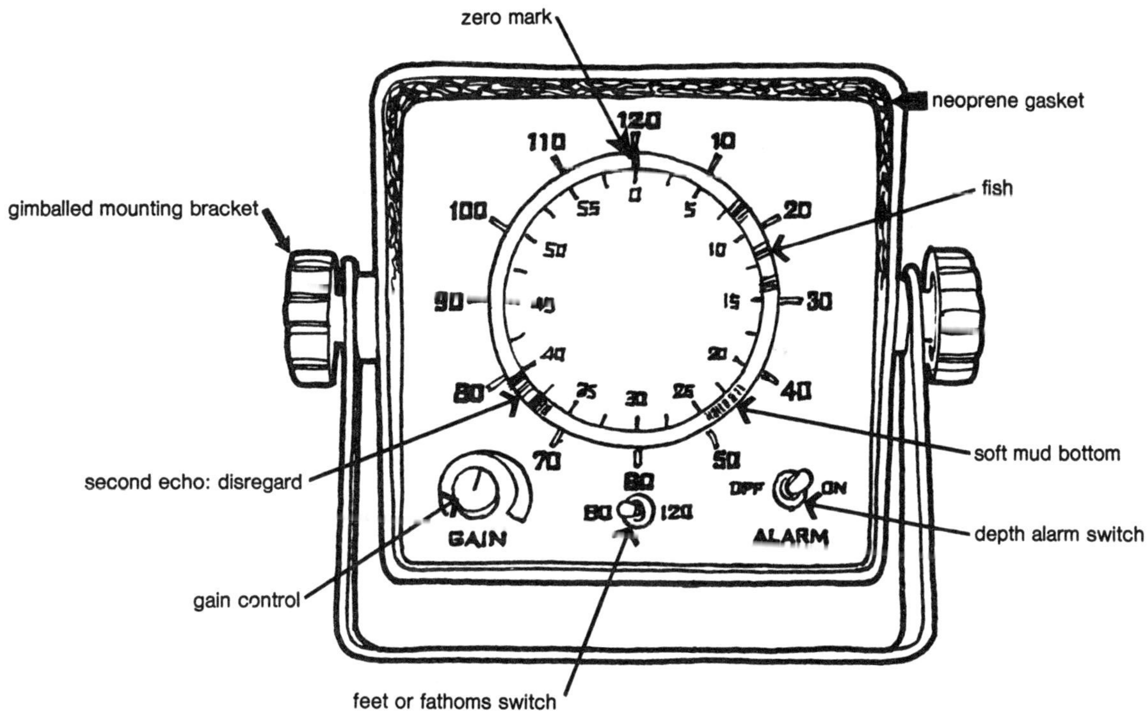

A flashing indicator is inexpensive and provides both bottom and fishfinding information. The price range with transducer is $100 to $500, and a typical size is 8½ inches high, 6½ inches wide, and 4 inches deep.

rocks, and an irregular bottom would probably show up as a very broad bottom flash that changes in width.

The latest innovation in flashers is a no-moving-parts liquid crystal display. The round dial remains, but the LCD black lines take over for the rotating disk and its mounted display lamp. The same LCD display can also give you (with separate sensors) water temperature and boat speed; digital depth displays and predetermined depth alarms are possible, too. Watch the LCD rotary-type display grow in popularity.

A flasher should have a gain control knob on the front panel with which to adjust the sensitivity of the receiver circuitry. Tuning this efficiently takes practice. Too much gain causes loss of target discrimination and may even cause false targets to appear; too little gain will cause weakening or loss of bottom echoes.

If you are really into serious fishfinding, you'll want a recording depthsounder, also known as a "fishfinder." There are two types—*chart recorders* and *video* or *CRT* (cathode ray tube) *recorders*. Chart recorders are excellent, and they are quite useful in navigating in deep or shallow waters. The chart paper will give you an accurate, permanent record of the bottom and everything in the water column. A typical reading might depict your downrigger being brought up from 70 feet for a lure change, large fish feeding off the bottom, game fish just below the surface of the water, and a well-defined wreck or obstruction on the ocean floor that might be a great candidate for skin diving exploration. Chart recorders will also assist you in following charted bottom contours to find your way home in the fog, and they will let you see the trend of a shoaling channel as you reenter a harbor.

Chart or graph recorders are also getting smarter, so if you're going to get a graph recorder, get a smart one. You can now adjust a dial to zoom in on the exact depth you want to examine. Want a closer look at that possible game fish at 40 fathoms or the ocean floor beneath you? Zoom in and take a long gander. And if you need to mark a spot on the chart where your downrigger hit a snag, push a button and there it is!

Many recorders feature bottom definition circuitry, variously referred to as "white line," "gray line," or by one of several other monikers. By purposely reducing bottom echo, this circuitry isolates fish that are floating just off the bottom. Some depth recorders will

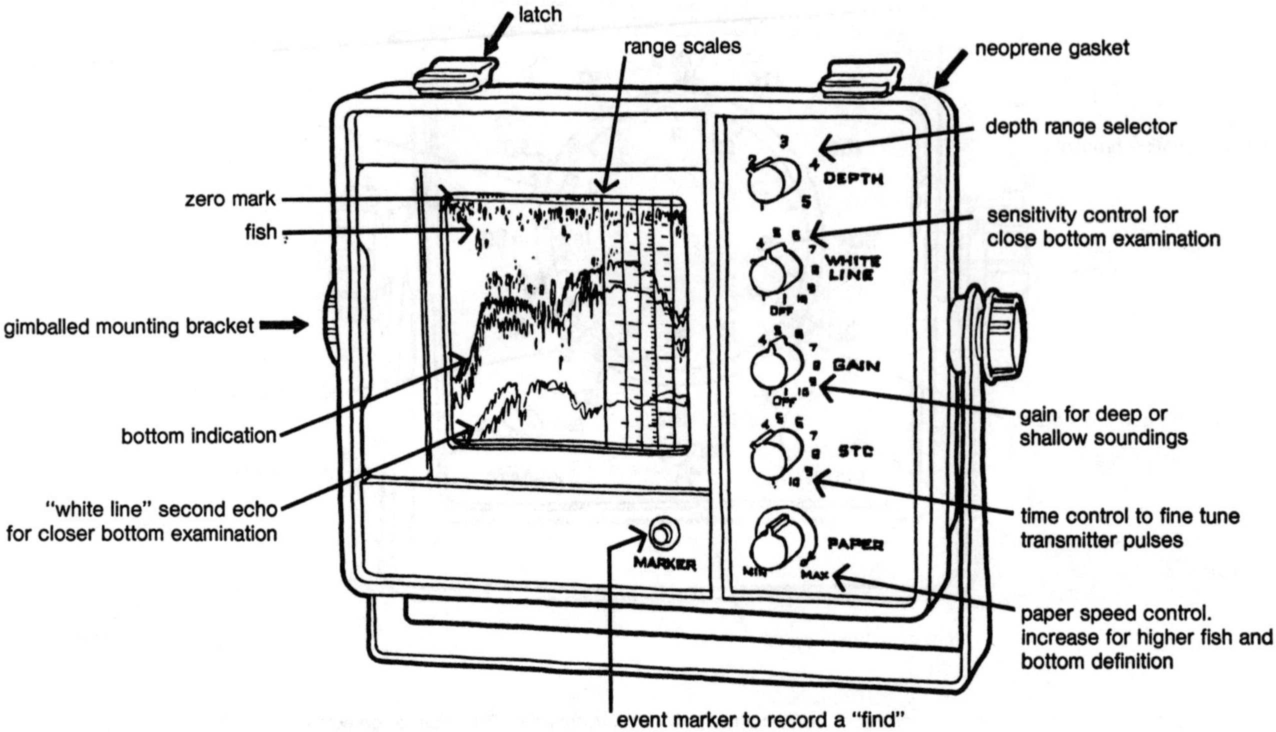

A chart recorder. This type of sounder is a favorite among fishermen, offering a tremendous amount of bottom and fish information. It is extremely useful in navigation as well as fishfinding. The price range with transducer(s) is $300 to $1,000, and a typical size is 9½ inches high, 12 inches wide, and 4 inches deep. Video sounders (not pictured in this chapter) are larger (12½ inches by 14 inches by 12 inches) and more expensive ($600 to $2,000).

allow you to change paper speed, too; a faster paper speed yields better bottom definition, "definition" implying the ability to detect small targets and deep bottoms (sensitivity) and to separate crowded targets (resolution). All of this is available in the more sophisticated depth recorders that offer variable paper speed, variable pulse length, and a combination of switchable transducers in addition to power and gain controls.

The latest innovation in depth recorders is the liquid crystal display, in which moving chart paper is replaced by black markings on an LCD display screen. These units are usually completely waterproof, and they'll let you see up to 32,000 bits of information at one time. You can recall one full screen of past information by pressing a single button, and an auxiliary tape recorder will allow you to record the digital information for later replay. The LCD readout is just as detailed as your digital watch, so you can imagine the precision of each target. The only characteristic you must get accustomed to is that every target is a miniature square, as opposed to a faithful reproduction of what the transducer is "seeing." An ascending bottom line that would appear smooth on a chart recorder will look like miniature ascending steps on the LCD readout. We should see better resolution in the future, as manufacturers put dots in every LCD frame.

Like chart recorders, video or CRT sounders present a picture of the bottom that moves from right to left. Rather than retaining the bottom imagery on chart paper or an LCD display, however, the video sounder memorizes it and displays it on a high voltage cathode ray tube. Some displays are black and white, and

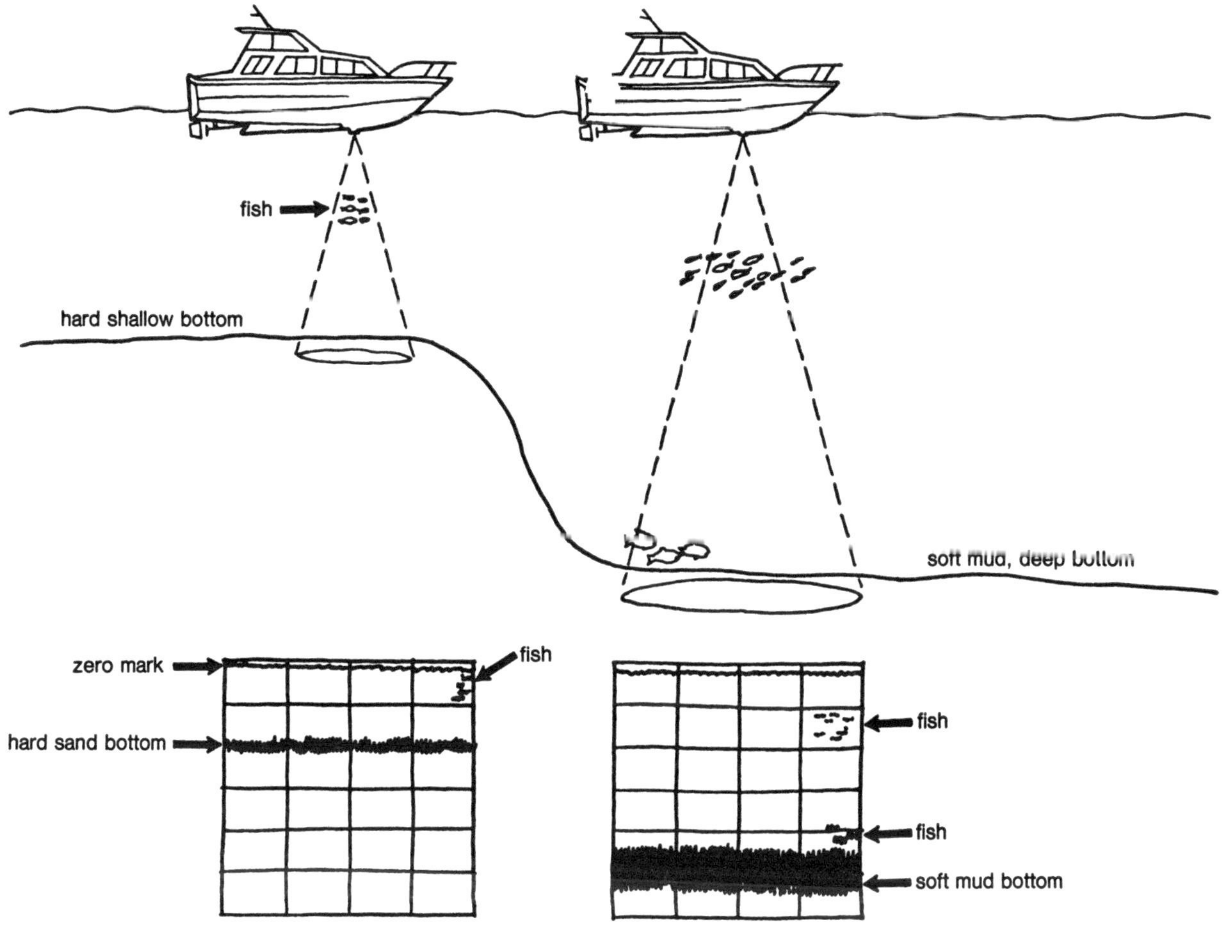

Vertical lines on a recording sounder's display represent time that has elapsed as the display moves from right to left, and the horizontal lines represent depths in feet, fathoms, or meters.

others are colored. These video screens should not be confused with true sonars, which "see" in all directions and paint a picture every second or so. Once the picture is "painted" on a video screen, it begins to move to the right; all updates occur at the right-hand side of the screen, with the remaining portion showing what has already taken place. Commercial fishermen have been using full-color video displays in which different colors represent different return echo intensities. It is possible to pick out a species of fish by the intensity and color of its return echo on the color video sounder. These units are usually only appreciated and purchased by serious fishermen.

Before you purchase a depthsounder, consider whether or not you need one that is waterproof, or at least weatherproof. Manufacturers are finally giving us completely waterproof sounders that can be hosed down after a day of fishing, and if you're going to be running your depthsounder in an open cockpit, then by all means select a unit that will take this type of punishment.

Installation of the indicator unit is straightforward, and you can accomplish it yourself. The units don't draw much current, and manufacturers usually suggest No. 10 wire for the positive and negative leads. This wire for voltage goes over to a switch panel, and there is nothing tricky about it—red to hot, and black to negative. Most sounders don't require an additional wire to ship's ground, but it's a good idea, helping to reduce static. A single No. 12 wire from the ground post on the sounder (or under any screw) should be run to your ship's ground plate or any common ground. Ground to your engine if nothing else is available.

Preventative maintenance involves cleaning the

power connection, the transducer plug, and other connections regularly, and making sure the unit is firmly in its mounting bracket. There is little to do inside the unit; sounders seldom require recalibration.

It is recommended that you paint the face of the transducer with antifouling paint. Usually one thin coat of bottom paint will keep the barnacles from forming on your transducer. Go with the bottom paint, not the barnacles. (If you should ever encounter growth of any kind on your transducer, take great care in removing it. That quartz crystal is protected only by a thin layer of plastic, and if the plastic gets gouged it skews the acoustical energy in different directions.)

Depthsounder malfunctions are usually traceable to the indicator unit, not the transducer, and the unit can easily be removed and sent back to the dealer or manufacturer for repair. Rarely do well-installed transducers go bad unless they have been physically damaged. If you treat your equipment properly, your depthsounder should work flawlessly for years.

Finally, let's talk about something you might not have known a sounder is capable of—and that's serious navigation. For example, you can tell what type of bottom you are over from a recorder, video recorder, or flashing indicator, though not from a digital. Over a sandy bottom, you will see a hard bottom indication, while kelp on the bottom will appear as a broad bottom signal. If you are cruising in unknown waters and following charted depth contours, remember that charts are based on mean low water. You can safely navigate around islands by following bottom contours and watching the chart. Unsure what island you are going around? Look for underwater hills and valleys by cruising in a straight line alongside the island, and match the depth configurations you see to the chart. In similar fashion, with a chart, dividers, a stop watch, and a little mathematics, it's possible to use a depthsounder for determining speed over the bottom. Watch those bottom contours—charts are quite detailed for this purpose. Your charted depth contour can also constitute a line of position to substantiate bearings from a radio direction finding station.

BUYING TIPS

If you've read this far, you probably have a fairly clear idea of what kind of depthsounder best meets your particular needs. Use the information in this chapter as a springboard for the examination of manufacturers' literature and advertising claims. When you have sifted through all of these, your chief remaining question is likely to be where to buy the unit.

A depthsounder is a logical piece of equipment to purchase from a catalog-discount house, because you're not giving up much in the way of service in order to save some money. Marine electronics dealers will install the indicator unit in your boat, but that part of the installation you can easily do yourself. They will *not* install the transducer, so you'll have to pay a boatyard to do that or struggle through it yourself in any case. Finally, if the indicator unit does malfunction, it will be easy to disconnect and ship directly to the manufacturer for repair.

3

VHF Radio

by Gordon West

Marine VHF radio enables you to communicate with other vessels up to 20 miles or more away, or with shore stations at greater distances. The VHF signals, having a short wavelength, are not much affected by local weather conditions, so VHF radio is a very reliable mode of marine communication. Before any other marine radio can legally be carried in your vessel, you must have a marine VHF transceiver aboard.

Virtually all VHF radios are manufactured in the Orient. State-of-the-art sets are synthesized—that is, they employ a single crystal that can be split by means of factory-programmed microprocessor circuits to transmit and receive on any existing marine VHF channel. The more traditional crystal sets cannot cover all the channels without employing multiple crystals, a more ponderous and expensive approach. Crystal sets are still being manufactured, but these generally offer only 12- or 24-channel coverage. Some 55-channel synthesized sets cost less than 24-channel crystal sets.

Armed with a basic understanding of the available VHF channels and a few simple rules to follow when choosing an antenna and a transceiver, you can put together a VHF station that suits your boat and satisfies your requirements. When you do, remember that perhaps the biggest problem facing marine VHF radio is the abuse of broadcasting protocol by pleasureboat operators. Mariners should learn and adhere to the standard marine VHF procedures.

THE CHANNELS

The marine VHF (very high frequency) band is centered around 156 megahertz (MHz), between Channels 6 and 7 on your television set. It is a group of frequencies standardized throughout the world by international agreement of the World Administrative Radio Conference (WARC). The band encompasses approximately seventy-eight 25-kHz channels between 156.050 MHz (Channel 1) and 157.425 MHz (Channel 88A) for ship transmitting; certain public telephone shore stations will also transmit on frequencies from 161.800 MHz (Channel 24 receive) to 162.025 MHz (Channel 88 receive, not applicable in the United States). The VHF band also includes several national weather transmitting frequencies centered around 162.550 MHz (weather channel 1 receive). Those frequencies between the ship transmit

Frequency Allocations

Channel 7 (Television)	179 MHz
Very High Frequency (VHF) Radiotelephone	156.0–162.55 MHz
EPIRB (Class C)	156.8 MHz
EPIRB (Classes A and B)	121.5 MHz
Omni/VOR	108.0–118.0 MHz
FM Radio	88–108 MHz
Channels 2–6 (Television)	54–88 MHz
Citizen's Band (CB) Radiotelephone	26.9–27.3 MHz
High Frequency Single Sideband Radiotelephone	4.0–23.0 MHz
Medium Frequency Single Sideband Radiotelephone	1.6–3.5 MHz (1,600–3,500 kHz)
AM Radio	530–1,600 kHz
Marine Radio Beacons	190–400 kHz
Loran-C	100 kHz
Decca	70–130 kHz
Omega	10.2–13.6 kHz

and shore transmit channels are allocated to other radio services (157.450 MHz to 161.595 MHz).

The number of channels that a mariner has access to in the United States is actually no greater than 55, some channels being paired differently with shoreside channels in other countries. Examples are channels 23A and 88A. The letter "A" after any VHF channel indicates that the channel has a split capability, being used in the United States for simplex (same frequency) transmitting and receiving, while having another use elsewhere. Channel 23A might be used by Coast Guard auxiliary members (with special permission from local Coast Guard units) to communicate with Coast Guard vessels. Channel 88A is used in almost all areas of the United States for ship-to-ship simplex commercial communications as well as for fish spotting by aircraft. In other countries, such as Canada, Channel 23 (without the "A" designation) is used for communications to a shoreside marine telephone service where the transmit and receive frequencies are different (a duplex split of 4.6 MHz). In the Great Lakes, mariners use Channel 88 (no "A") in the duplex mode to place telephone calls with shore stations that transmit on a different frequency (4.6 MHz higher) than the ship stations.

Modern VHF sets have programmed microprocessor circuits that tell them on what channels to operate simplex and when to switch automatically to duplex (different receive). Since on the "A" channels (and only on the "A" channels), this may vary by country, sets are usually factory programmed for use either in the United States—designated as the "United States" or "domestic" mode—or elsewhere—the "foreign" mode. A few sets offer a manual override for switching modes. Apart from these few exceptions, no matter where you cruise throughout the world, all VHF channels are the same.

The most important channel is 156.800 MHz, Channel 16, the international distress and calling frequency. Every VHF radio must have this channel installed in it, and it must be monitored at all times when your set is turned on, except when you are operating on another frequency. The United States Coast Guard and other rescue agencies worldwide monitor this channel 24 hours a day, and it is also monitored by orbiting search and rescue satellites that will retransmit the distinctive sounds of an emergency position indicating radio beacon (EPIRB) signal, along with its approximate location, when one is received. This channel must only be used for distress and safety calls or establishing communications with another vessel before switching to a working channel. You are not allowed to talk more than 59 seconds on VHF Channel 16 unless you are making an emergency communication.

VHF Channel 6 is the safety channel, to be used only for intership communications relating directly to the safety of vessels. Use this channel to communicate priority or urgent traffic to another vessel. Every VHF set must have Channel 6.

There are seven pleasureboat ship-to-ship communications channels: 9, 68, 69, 70, 71, 72, and 78. Use any of these to communicate with other boats in your area. The messages you transmit must relate directly to the operation of your vessel, and all communications must be kept to a minimum, since other vessels in your area may wish to use the same frequencies. Yacht clubs, ship supply stores, and fuel docks may also use Channels 9, 68, 69, 71, and 78 to communicate with ships.

SUGGESTED CHANNEL DISTRIBUTIONS FOR PLEASURE AND COMMERCIAL CRAFT

Channel Numbers	Type of Communication	Suggested Channel Selection		
		Recreational Vessels		Commercial Vessels
		6 ch.	12 ch.	12 chs.
16	DISTRESS, SAFETY & CALLING Intership & ship to coast	*	*	*
6	INTERSHIP SAFETY Intership. NOT to be used for non-safety intership communications	*	*	*
22A	Communications with U.S. Coast Guard ship, coast, or aircraft stations	1	1	1
1,5,65A, 66A,12,73, 14,74,63, 20,77	PORT OPERATIONS Intership & ship to coast		1	2
13,67	NAVIGATIONAL Intership & ship to coast		1	1
68,9,69, 71,78A,72	NON-COMMERCIAL Intership & ship to coast	1	3	
70,72	NON-COMMERCIAL Intership		2	
1,7A,9,10 11,18A,19A, 79A,80A,63	COMMERCIAL Intership & ship to coast			3
67,8,77, 88A	COMMERCIAL Intership			1
24,84,25, 85,26,86, 27,87,28 ***88	PUBLIC CORRESPONDENCE Ship to public coast	2	2	2
162.400 (WX-2) 162.550 (WX-1) 162.475 (WX-3)	NOAA WEATHER SERVICE Ship receive only	**	**	**

*These channels are REQUIRED to be installed in every ship station equipped with a VHF radio.

**The weather receive channels are half-channels (receive only) one or more of which are recommended to be installed in each ship station. Many manufacturers include one or more of these channels in their sets in addition to the normal six or twelve channel capacity.

***See limitations in table on page 20.

Pleasureboat operators are not allowed to use commercial frequencies, nor may commercial vessels use noncommercial frequencies. Of course, while it's illegal for pleasureboat operators to transmit on Channel 88A, for example, there is no harm in monitoring this channel to get the latest scoop from commercial fish spotters! Marine Channel 9 is shared by both commercial and pleasure boats, so if you wish to talk to a commercial fishing boat, first call him on 16 and then switch to 9.

When you hear the distinctive warbling sound of an emergency position indicating radio beacon on Channel 16, it means that someone in your area is in distress. The Class C EPIRB* will alternate automatically between Channel 16 and the adjacent Channel 15. EPIRBs are only one-way transmitters, so you won't be able to establish communications with the vessel in trouble. Your role is immediately to contact the Coast Guard on Channel 16 and let them know that you hear an EPIRB in your area. Call the Coast Guard when the EPIRB switches from Channel 16 to Channel 15 so you can hear their response. They may ask whether you have direction finding capability, and if so, what bearing the EPIRB is from your computed location.

The table lists several channels used for placing phone calls to marine telephone operators ashore. It's best to register ahead of time with your local marine operator before you attempt to place a phone call through their service. This is easily accomplished with a shoreside call. Marine electronics specialty houses may also have marine operator registration forms for specialized phone services not listed with your AT&T operator. Many of these offer terrific telephone service, including message handling and conference calling. A marine telephone service is a great bargain. There is no monthly fee if you don't use their service. You're only charged for the phone calls you place, and if you sit the winter out and never make a call, you are not billed. The average price of a three-minute call to a local shoreside station is generally less than $2. Some phone companies may charge a one-time registration fee or a small yearly renewal fee.

Some marine telephone operators announce traffic lists on VHF Channel 16 on the hour. Due to tremendous traffic on Channel 16 during the summer months, however, you may wish to monitor, with a second VHF set, your marine telephone channel to receive phone calls the second they are placed to your

**Class C EPIRBs, the usual choice for coastwise vessels, transmit at 156.8 MHz. Class A and B EPIRBs for high-seas use transmit at 121.5 MHz.*

MARINE VHF CHANNEL ALLOCATIONS

Channel Number	Ship Transmit (MHz)	Ship Receive (MHz)	Channel Usage — Intended Use
1 3 63	156.050 156.150 156.175	156.050 156.150 156.175	PORT OPERATIONS AND COMMERCIAL (Intership and Ship-to-Coast). Available for use within the U.S.C.G. designated Vessel Traffic Services (VTS) area of New Orleans, and the Lower Mississippi River.
5	156.250	156.250	PORT OPERATIONS (Intership and Ship-to-Coast). Available for use within the U.S.C.G. Vessel Traffic Services radio protection areas of New Orleans and Houston.
6	156.300	156.300	INTERSHIP SAFETY. Required for all VHF-FM equipped vessels. For intership safety purposes and search and rescue (SAR) communications with ships and aircraft of the U.S. Coast Guard. Must not be used for non-safety communications.
7A	156.350	156.350	COMMERCIAL (Intership and ship-to-coast). A working channel for commercial vessels to fulfill a wide scope of business and operational needs.
8	156.400	156.400	COMMERCIAL (Intership). Same as channel 7A except limited to intership communications.
9	156.450	156.450	COMMERCIAL AND NON-COMMERCIAL (Intership and ship-to-coast). Some examples of use are communications with commercial marinas and public docks to obtain supplies or schedule repairs and contacting commercial vessels about matters of common concern.
10	156.500	156.500	COMMERCIAL (Intership and ship-to-coast). Same as channel 7A.
11	156.550	156.550	COMMERCIAL (Intership and ship-to-coast). Same as channel 7A. It should be noted, however, in certain Ports channels 11, 12 and 14 are to be used selectively for the Vessel Traffic Service now being developed by the United States Coast Guard.
12	156.600	156.600	PORT OPERATIONS (Intership and ship-to-coast). Available to all vessels. This is a traffic advisory channel for use by agencies directing the movement of vessels in or near ports, locks, or waterways. Messages are restricted to the operational handling, movement and safety of ships and, in emergency, to the safety of persons. It should be noted, however, in certain ports 11, 12 and 14 are to be used selectively for the Vessel Traffic Service being developed by the United States Coast Guard.
13	156.650	156.650	NAVIGATIONAL—(Ship's) bridge to (Ship's) Bridge. This channel is available to all vessels and is required on large passenger and commercial vessels (including many tugs). Use is limited to navigational communications such as in meeting and passing situations. Abbreviated operating procedures (call signs omitted) and 1 watt maximum power (except in certain special instances) are used on this channel for both calling and working. For recreational vessels, this channel should be used for listening to determine the intentions of large vessels. This is also the primary channel used at locks and bridges.
14	156.700	156.700	PORT OPERATIONS (Intership and ship-to-coast). Same as channel 12.
15		156.750	ENVIRONMENTAL (Receive only). A receive only channel used to broadcast environmental information to ships such as weather, sea conditions, time signals for navigation, notices to mariners, etc. Most of this information is also broadcast on the weather (WX) channels.
16	156.800	156.800	DISTRESS, SAFETY, AND CALLING (Intership and ship-to-coast). Required channel for all VHF-FM equipped vessels. Must be monitored at all times station is in operation (except when actually communicating on another channel). This channel is monitored, also, by the Coast Guard, public coast stations and many limited coast stations. Calls to other vessels are normally initiated on this channel. Then, except in an emergency, you must switch to a working channel.
17	156.850	156.850	STATE CONTROL. Available to all vessels to communicate with ships and coast stations operated by state or local governments. Messages are restricted to regulation and control, or rendering assistance. Use of low power (1 watt) setting is required by international treaty.
18A	156.900	156.900	COMMERCIAL (Intership and ship-to-coast). Same as channel 7A.
19A	156.950	156.950	COMMERCIAL (Intership and ship-to-coast). Same as channel 7A.
20	157.000	161.600	PORT OPERATIONS (Ship-to-coast). Available to all vessels. This is a traffic advisory channel for use by agencies directing the movement of vessels in or near ports, locks, or waterways. Messages are restricted to the operational handling, movement and safety of ships and, in emergency, to the safety of persons.
21A	157.050	157.050	U.S. GOVERNMENT ONLY.
22A	157.100	157.100	COAST GUARD LIAISON. This channel is used for communications with U.S. Coast Guard ship, coast and aircraft stations after first establishing communications on channel 16. Navigational warnings and, where not available on WX channels, Marine Weather forecasts are made on this frequency. *It is strongly recommended that every VHF radiotelephone include this channel.*
23A	157.150	157.150	U.S. GOVERNMENT ONLY.
24	157.200	161.800	PUBLIC CORRESPONDENCE (Ship-to-coast). Available to all vessels to communicate with public coast stations. Channels 26 and 28 are the primary public correspondence channels and therefore become the first choice for the cruising vessel having limited channel capacity.
25	157.250	161.850	PUBLIC CORRESPONDENCE (Ship-to-coast). Same as channel 24.
26	157.300	161.900	PUBLIC CORRESPONDENCE (Ship-to-coast). Same as channel 24.
27	157.350	161.950	PUBLIC CORRESPONDENCE (Ship-to-coast). Same as channel 24.
28	157.400	162.000	PUBLIC CORRESPONDENCE (Ship-to-coast). Same as channel 24.
65A	156.275	156.275	PORT OPERATIONS (Intership and ship-to-coast). Same as channel 12.
66A	156.325	156.325	PORT OPERATIONS (Intership and ship-to-coast). Same as channel 12.
67	156.375	156.375	COMMERCIAL (Intership). Same as channel 7A except limited to intership communications.
68	156.425	156.425	NON-COMMERCIAL (Intership and ship-to-coast). A working channel for non-commercial vessels. May be used for obtaining supplies, scheduling repairs, berthing and accommodations, etc. from yacht clubs or marinas, and intership operational communications such as piloting or arranging for rendezvous with other vessels. It should be noted that channel 68 (and channel 70 for intership only) is the most popular non-commercial channel and therefore is the first choice for vessels having limited channel capacity.
69	156.475	156.475	NON-COMMERCIAL (Intership and ship-to-coast). Same as channel 68.
70	156.525	156.525	NON-COMMERCIAL (Intership). Same as channel 68, except limited to intership communications.
71	156.575	156.575	NON-COMMERCIAL (Intership and ship-to-coast). Same as channel 68.
72	156.625	156.625	NON-COMMERCIAL (Intership). Same as channel 68, except limited to intership communications.
73	156.675	156.675	PORT OPERATIONS (Intership and ship-to-coast). Same as channel 12.

Frequencies (MHz)			Channel Usage
Channel Number	Ship Transmit	Ship Receive	Intended Use
74	156.725	156.725	PORT OPERATIONS (Intership and ship-to-coast). Same as channel 12.
77	156.875	156.875	COMMERCIAL (Intership). Same as channel 7A except limited to intership communications.
78A	156.925	156.925	NON-COMMERCIAL (Intership and ship-to-coast). Same as channel 68.
79A	156.975	156.975	COMMERCIAL (Intership and ship-to-coast). Same as channel 7A.
80A	157.025	157.025	COMMERCIAL (Intership and ship-to-coast). Same as channel 7A.
81A	157.075	157.075	U.S. GOVERNMENT ONLY.
82A	157.125	157.125	U.S. GOVERNMENT ONLY.
83A	157.175	157.175	U.S. GOVERNMENT ONLY.
84	157.225	161.825	PUBLIC CORRESPONDENCE (Ship-to-coast). Same as channel 24.
85	157.275	161.875	PUBLIC CORRESPONDENCE (Ship-to-coast). Same as channel 24.
86	157.325	161.925	PUBLIC CORRESPONDENCE (Ship-to-coast). Same as channel 24.
87	157.375	161.975	PUBLIC CORRESPONDENCE (Ship-to-coast). Same as channel 24.
88A	157.425	157.425	COMMERCIAL (INTERSHIP). Same as channel 7A, except limited to intership communications and between commercial fishing vessels and associated aircraft while engaged in commercial fishing.
WX1	—	162.550	WEATHER (RECEIVE ONLY). To receive weather broadcasts of the Department of Commerce, National Oceanic and Atmospheric Administration (NOAA).
WX2	—	162.400	WEATHER (RECEIVE ONLY). Same as WX1.
WX3	—	162.475	WEATHER (RECEIVE ONLY). Same as WX1.

NOTE. The addition of the letter "A" to the channel number indicates that the ship receive channel used in the United States is different from the one used by vessels and coast stations of other countries. Vessels equipped for U.S. operations only will experience difficulty communicating with foreign ships and coast stations on these channels.

vessel from a shoreside station. You can do this with a small handheld transceiver.

Since telephone channels are separate transmit and receive frequencies, it's impossible to transmit ship-to-ship on vacant phone channels. You simply won't hear the other ship station respond because your receiver is tuned 4.6 kHz higher than your transmitter. If you wish to establish ship-to-ship communications through a telephone operator, however, because the two of you are well beyond VHF line-of-sight range, the phone company will patch you in together. Your signals will boom in loud and clear thanks to the phone operator "repeater." You could be as much as 1,000 miles apart yet still stay in contact. There is a charge for the service.

The other channels are described briefly in the accompanying table and are explained more fully in VHF owners' manuals, in piloting books, and in the *Marine Radiotelephone User's Handbook*, put together by the Radio Technical Commission for Marine Services in cooperation with the FCC. A copy can be purchased for $7.95 from the RTCM, P.O. Box 19087, Washington, D.C. 20036.

RANGE

VHF radio waves travel over the water on a line-of-sight basis plus a mere 22 percent over-the-horizon allowance for "super-refraction"; that is, the VHF waves, being only about 2 meters long, will not bend around the horizon to any great extent. Ultimate range is determined by the height of your antenna over the water and by its gain. The Coast Guard and marine operators often use well-elevated shoreside antenna locations to offer excellent range to ship stations. (They also frequently use remote receivers that pick up distant calls.) The usual range to any Coast Guard station or marine operator is in excess of 50 miles from any good VHF 25-watt station. If you use a handheld VHF, your range may diminish to a maximum of 35 miles, but output power on a clear channel is not of primary importance to the ultimate range of a VHF signal.

The maximum range of ship-to-ship communications is substantially less because neither station has any appreciable height over the water. Average ship-to-ship communications are limited to 15 to 20 miles on most days. Given the heights of both antennas, specific maximum ranges are easily calculated: An antenna 10 feet off the water "sees" the horizon at 4 miles, and two such antennas will "see" each other at 8 miles. Adding the 22 percent factor for refraction of VHF radio waves over seawater gives us a minimum communications distance of 10 miles. Going to higher power levels will add no more than a half mile or so.

You might occasionally establish ship-to-ship communications between vessels up to 50 miles apart as a result of "tropospheric ducting." This most often occurs when a high-pressure system slips over your local cruising area, and there is little wind. A temperature inversion forms at an altitude of approximately 1,000 feet, trapping VHF radio signals below it. In Southern California, the "smog belt" at 1,000 feet sometimes carries VHF signals up to 300 miles away, ship-to-ship. These are not "skip" signals—rather, they are signals experiencing an intense super-refraction. Communications with shoreside stations over 400 miles away may be possible on rare occasions.

ANTENNAS

The most important element of good range is your antenna system. Every radio needs a good ground, and VHF antennas have the ground network already built in, which is why they are called antenna systems.

The 3-foot white fiberglass or stainless steel whip, the most common antenna for sailboats, is fine for local communications and small enough to be out of the way. It should be mounted as high as possible, however. Sailboat owners choosing a masthead mount for maximum range should use a stainless steel base-loaded whip. Although you *could* put a 3-foot white fiberglass whip on the masthead, the stainless steel antenna system will offer less windage with equal performance and will be almost undetectable on its lofty perch.

VHF antennas are normally rated by the amount of gain they exhibit over an isotropic antenna, a theoretical antenna with zero gain. The 3-foot whip is referred to as having a 3 decibel gain, which represents a twofold power increase over the imaginary antenna on both transmit and receive. The 9-foot white fiberglass antenna system is usually rated at 6 dB gain, that is, a fourfold power increase, and it is a favorite among small powerboat operators. It perches atop a cabin with only a bottom mount. Sailboat enthusiasts might use a 9-foot whip off the stern in lieu of the 3-foot whip at the masthead, but the latter will give and receive better signals, height being more important than decibel gain.

The granddaddy of them all is the 21-foot white fiberglass antenna rated at 9 dB gain. These antennas require a second plastic support bracket, and they appear almost exclusively on larger powerboats, where they are attached to the side of the wheelhouse. Commercial fishermen use 9 dB antennas high atop their fishing towers for maximum range. You may hear tales about antenna pattern distortion in heavy seas with a 9 dB gain antenna, since the higher gain concentrates transmitted signals and receiver sensitivity in a comparatively thin plane perpendicular to the antenna. Be assured, however, that no signal can top that from a 9 dB gain antenna regardless of how much you are pitching or rolling. This antenna will transmit better, receive better, and outperform any smaller antenna hands down, regardless of sea conditions. For the ultimate in range, the 9 dB gain antenna is the way to go if your boat is large enough to take it.

Select only name-brand VHF antennas for a lifetime of operation—bargain brands may fracture on the inside and give dismal results after a few years at sea. The conductors of cheap antennas may be a fragile network of stripped coax cable, as opposed to the rugged soldered tube of a good antenna. The cheap antenna is unlikely to have the physical integrity to withstand constant vibration.

The future trend may well be toward antennas that combine several shipboard functions, such as VHF radio, loran, and SatNav. One manufacturer has anticipated this trend with their combination antenna for loran and VHF. It is compatible with any VHF transceiver, but only with their own loran unit.

COAXIAL CABLE

A marine VHF antenna is connected to the transceiver using coaxial cable. Some antennas have the coax cable built in, this cable being adequate for most short hookups. If a longer cable is needed, however (such as when an antenna is intended for masthead use), no cable is built in. For moderately long runs, use RG-8/X (or RG-8X) coaxial cable, which offers considerably less signal loss than the smaller RG-58AU (or RG-58A/U) cable. If your VHF antenna run is long and you want the ultimate in performance, select RG-213/U cable for minimum losses at 156 MHz. The commercial boat operator may go one step further and put in the semirigid Belden 9913 coaxial cable, which offers almost no loss at all over a 100-foot run at 156 MHz.

A coaxial cable is a shielded transmission line in which one conductor is mounted coaxially inside the other. Think of it as a rod-within-a-sleeve configuration. Due to skin effect, radio frequency current is carried on the outside surface of the inner conductor and the inside surface of the outer conductor. This keeps the energy within the cable from escaping except at the ends, where you want it. Generally, the larger (and more expensive) coaxial cable will offer radio energy less resistance with less loss than will smaller cable. This is simply because there is more cross-sectional area of conductor material to carry the current. Conductor losses are a direct function of the conductivity of the conductors, too, with most coaxial cables consisting of copper or copper and aluminum.

Coaxial cable impedance (usually 50 ohms) is a function of the ratio of the diameters of the inner and outer conductors. Most marine radio transceivers have an output of 50 ohms and are compatible with 50-to-52-ohms-impedance coax cable. This immediately rules out using discarded cable TV coax (72 ohms), including that huge CATV coax cable that a friend snipped from a coil left outside by the local cable TV company.

Another important consideration for coax cable in a marine environment is its moistureproof capabilities. Coaxial cable jackets made of plastic alone will eventually leak moisture, but "noncontaminating" jackets made of polyvinyl chloride will resist the threats of moisture provided the cable ends have been properly terminated and sealed.

The inner dielectric (nonconducting element) can also provide a nice home for moisture, and both foam-core and air-core coaxial cables will allow moisture, through capillary action, to travel up and down the inner core and the outside copper braid. Noncontaminating coaxial cable with a solid PVC dielectric will keep moisture from getting to the center conductor and will help decrease the capillary action of moisture along the outside braid. There are some coaxial cables that not only have a solid dielectric but are also backfilled with a sticky "goo" that completely covers

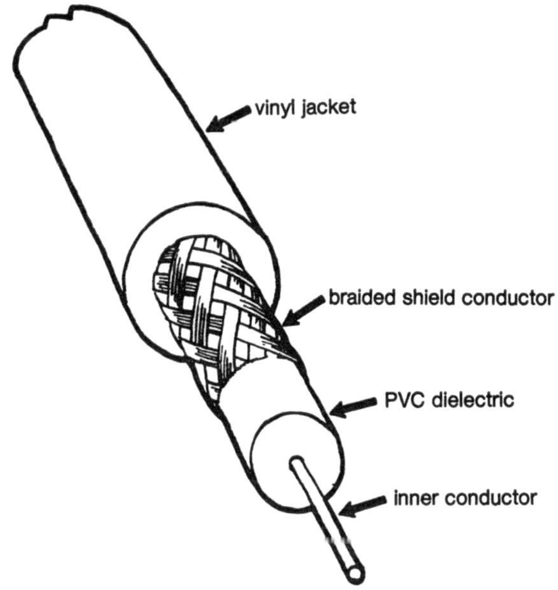

Cut-away view of a coaxial cable.

the outside braid within the PVC jacket to repel moisture. While this may be the ultimate coax, it's messy to work with and pretty hard to find these days.

Purchase coaxial cable only from reputable ham radio dealers—CB-type coax is out of the question. Although CB radio coaxial cable may carry the "RG" (Radio Government) specification, the CB cable manufacturers produce a cheapened version of this coax that doesn't actually meet the original government specs. They skimp on the number of individual center conductors as well as the number of individual conductors in the braid of the copper shield. Most CB coax is quite leaky because of the huge gaps in the braid, and it distorts signals and causes impedance mismatches when the cable is flexed to any extent.

RG-8/X cable is a good size for interconnecting cables, jumpers, and short mobile runs between a handheld transceiver and a small, solid-state, linear amplifier.* Good quality RG-8/X (sometimes called

---

**The use of a linear amplifier to boost VHF output above 25 watts is illegal. A linear amplifier may be used to increase the output of a handheld unit to a maximum of 25 watts, but there are associated problems of power drain and overheating, and the cost of a handheld unit plus linear amplifier may exceed the cost of a 25-watt VHF set.*

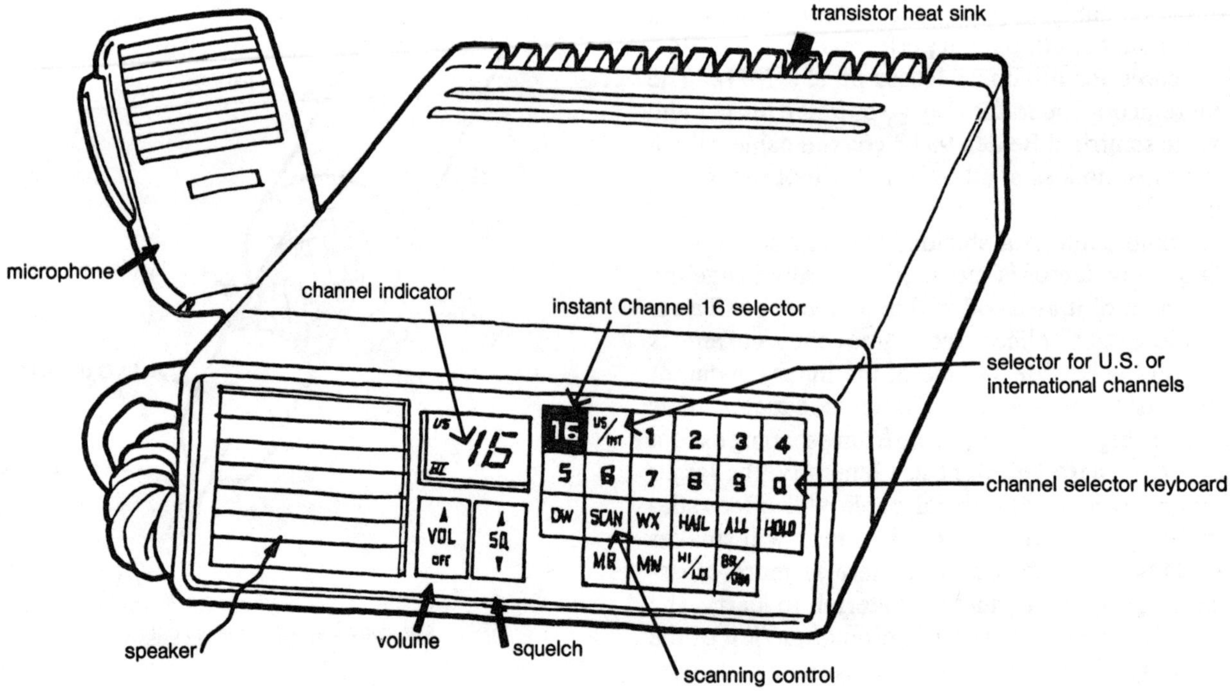

There is little standardization of display and keypad labels among keyboard VHF sets. This unit includes an automatic channel-scanning capability, connection to a loud hailer, weather frequencies, and a dimmer control for night illumination. All VHF radios must also have a switch or keypad button for low and high power output. The price range for a synthesized keyboard unit is $300 to $500, plus another $30 to $40 for the antenna. A typical size is 6 inches wide, 10 inches deep, and 3 inches high.

miniature RG-8/U) takes the regular PL-259 antenna connector, but the slightly larger diameter UG-176/U reducer. This reducer is as common as apple pie, and you can purchase it anywhere. For a run over 20 feet, a better choice, as mentioned, is the larger coaxial cable called RG-213/U. This is exactly the same size as RG-8/U, but the designation RG-213/U signifies a type IIA noncontaminating, plasticized, synthetic-resin, protective, black jacket that will resist abrasion and will not be damaged by ultraviolet sunlight rays. You can run this cable in the bilge or submerge it in water without a chance of contamination or leakage, and it should last a minimum of 20 years. Coaxial cable without this special jacket and PVC interior may only last two years—another point against CB-type feedlines. The larger RG-213/U is almost twice as expensive as RG-8/X (59 cents versus 29 cents per foot in 1984), but it's worth the price. At 156.8 MHz, a 100-foot run of RG-213/U causes only about 2.3 dB loss.

TRANSCEIVERS

The pleasure craft operator may use fewer than 30 marine VHF channels for transmitting, but there's no real reason not to purchase a 55-channel synthesized VHF set that will tune in every United States channel plus a few Canadian channels at a flick of the switch. You don't pay much more for this capability. If you buy a set with 78 channels, on the other hand, you aren't getting any more U.S. channels—just a few European telephone channel capabilities. Shop accordingly. When a set boasts "101 channels," the manufacturer is using an advertising ploy; give one black mark for hyperbole.

All marine VHF sets offer 25 watts output, the maximum allowed by the FCC. In fact, thanks to imported power transistors, 25 watts is just loafing for most sets. In addition, all nationally advertised marine VHF transmitters must be accepted under Part 83 of the FCC rules and regulations. These requirements are stringent, so there is not much difference between one transmitter and another. If the set is designed for permanent installation, it will put out a clean 25 watts with good transmit audio.

Unlike the transmitter section, the receiver section of marine 25-watt VHF sets varies from extremely poor to extremely good. Manufacturers can cut costs in a VHF set by giving you an inefficient receiver. Try the set out at the electronics display of a dealer, and listen for adequate audio. Crank the volume all the way up and see whether or not it's really ear-splitting. It will need to be if you are in an open cockpit and you need to hear the set over the din of your engine at full throttle. The unit may be advertised as having 4 watts audio power, but these specs mean little. Hearing is believing. Check the channel readout—can you see the digits in the daylight? Dull light-emitting diodes (LEDs) are fast being replaced by the new bold liquid crystal displays (LCDs) that can be seen day or night.

It's pretty hard to determine receiver sensitivity as well as channel selectivity, and again, the specifications that accompany each VHF set are usually nothing more than copies of specs from their competitors. Sensitivity is the measure of the receiver's ability to hear weak signals, and you may read, for example, that a particular receiver is sensitive to a 0.35 microvolt signal at 20 dB quieting, or a 0.25 microvolt signal at 12 dB signal noise and distortion (SINAD). The greater the number of microvolts, the less the sensitivity of the receiver. Much more to the point, remember simply that sensitivity is easy to achieve, and most sets are sensitive regardless of price. Selectivity (the ability to reject adjacent channel frequencies) and intermodulation rejection (the ability to reject "invading" signals formed from the mixing of out-of-band frequencies) are the critical factors in receiver quality, and unfortunately, a good way to judge these is by the transceiver's price tag. Inexpensive VHF sets may allow out-of-band signals to come through on Channel 16, so instead of hearing the Coast Guard when you're in Boston Harbor, you hear the local towing company. Quite common with cheap sets. The meaningless, copied specifications may state that both selectivity and intermodulation rejection are higher than 60 dB, but it is better to turn on the radio at the showroom and have it hooked up to the biggest VHF antenna they have on the roof. Listen to Channel 16, and if you hear strange beeps, moans, groans, and tones that come and go, chances are the receiver is experiencing intermodulation and poor out-of-band signal rejection.

The squelch circuit is another indication of quality. Try this simple test—turn the squelch off so that background noise comes through the loudspeaker at a reasonable level. Now slowly advance the squelch until the background noise disappears. Did the noise stop abruptly or did it slowly fade away as the squelch was advanced? Abrupt squelch circuits cause a DC voltage gate to switch the receiver on and off—each time with an accompanying "pop"—as signals come and go in the squelch. You can hear this "pop" even when the volume is turned all the way down by listening carefully to the loudspeaker. It is not an indication of a good, sensitive VHF set. The better receiver and squelch circuit will gradually filter the background noise as the squelch goes into its silent mode. Extremely weak signals won't pop in and out of the squelch, but will appear gracefully as smooth audio coming through your loudspeaker. Instead of a crackling noise on a weak signal, you will hear the distant voices quietly emerge.

Except in crystal sets, the transceiver is frequency selected by the synthesizer within your VHF set. You have a choice of keyboard channel entry or the more traditional rotating knob. The keyboard-entry VHF sets have more features, such as memorization of channels (just keyboard-enter the channels you select), channel scanning (the set monitors by turn each preselected channel, stopping on an incoming signal), channel searching (the set searches *all* the factory-installed channels), and priority channel operation (the set remains on the working channel but scans the priority channel at brief intervals, locking there if it receives a transmission). You can preprogram some units to operate at high power on certain channels and low power on others*; a few units offer a hailer-speaker accessory, which substitutes for a foghorn and can be used as an intercom.

Rotating-knob units sit on one channel unless manually adjusted. Manufacturers are producing fewer of these because of this limitation and because the mechanical contacts may become fouled or corroded in time. The keyboard units substitute

**Marine VHF units are legally required to offer the option of high power (25 watts) or low power output, but in truth, most mariners tend to leave their sets at high power even when talking to nearby vessels.*

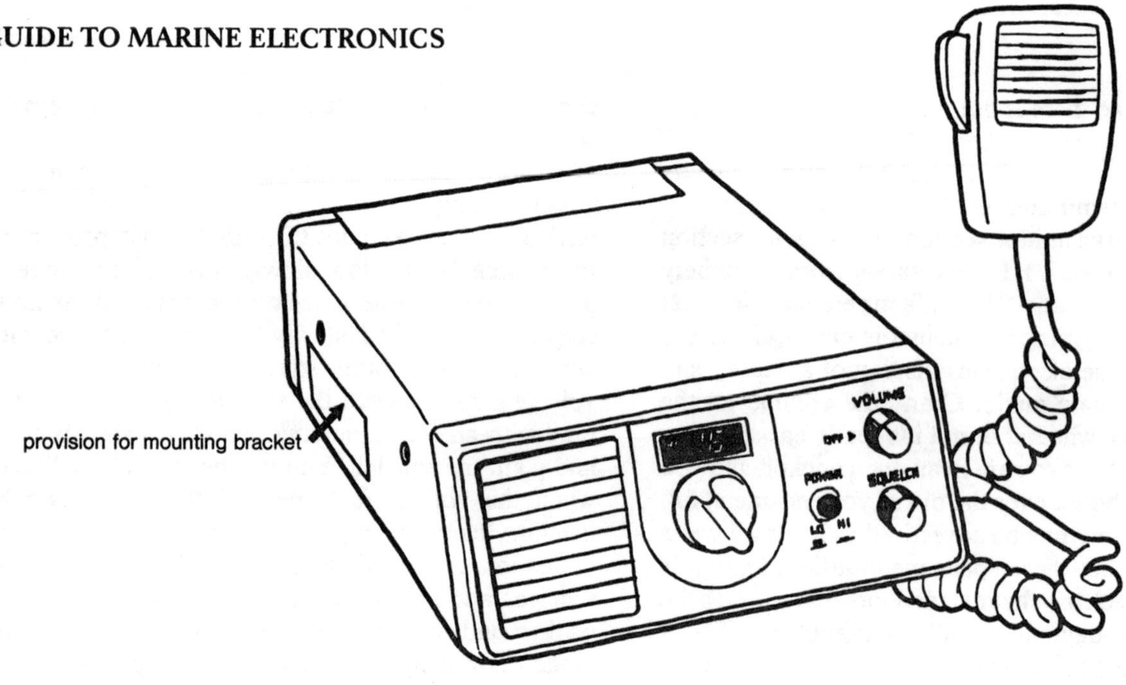

The simplest of VHF radios uses a rotary channel selector, but some such units have no more than 12 to 24 channels. That number is adequate for most boat owners unless they are cruising in international waters, where channels are used that are not authorized by the FCC for U.S. waters. Above the rotary channel selector on the unit shown here is the channel indicator, and to the right is the button for low/high power output. Farther right are the on/off/volume and squelch controls. The price range is $200 to $600 (without antenna), and a typical size is 6 inches wide, 10 inches deep, and 3 inches high.

miniature microswitches or pressure-sensitive membranes for mechanical contacts. Rotating-knob sets are still preferred by some mariners, however. Fishermen find the knobs easier to manipulate with gloved fingers than the small pads of a keyboard set, and a channel change can be accomplished without looking by rotating the knob through the requisite number of clicks from the previous, known channel. There are some excellent rotating-knob sets on the market.

If you plan to expose a keyboard unit to the weather, make sure the keyboard is covered by a weatherproof membrane. Some rotating-knob VHF sets are fully waterproof and can be completely hosed down without adverse effects. Behind each knob is a neoprene gasket that seals moisture out. If a radio is weatherproof or waterproof, you can be sure that the advertising literature will make a point of this. If the feature isn't mentioned in the literature, keep the set bone dry.

INSTALLATION

Mount the antenna as high as possible and at least three feet away from metal spars and standing rigging except when mounted atop these. Never clamp it to stainless steel stanchions or canopy frames, since these metal objects will absorb some of the signal.

It will do no harm to run the coaxial cable with other wiring. Since the signal is kept inside the coax by the outer braid, there should be no pickup of unwanted noise. If you have excess cable, chop it off. An unnecessarily long coil of coax behind the set will only rob you of power output and weaken signal receive capabilities. If you have never soldered a coaxial cable connector (PL-259, with or without the UG-255 adapter or reducer), let someone qualified do it for you. The coaxial cable connector improperly soldered, whether at the back of the radio or the base of the antenna, is a very common failure in most new installations. As an alternative, use a crimped, no-solder connector. These are available anywhere antennas are sold, and more and more manufacturers are starting to include them with their VHF radios and Loran and SatNav receivers, knowing that owner installation of this gear is on the increase. These connectors, which are simple to install, work very well but do not provide the longevity of a soldered connection. Still, they typically last about three years before the signal becomes intermittent and the connector must be replaced.

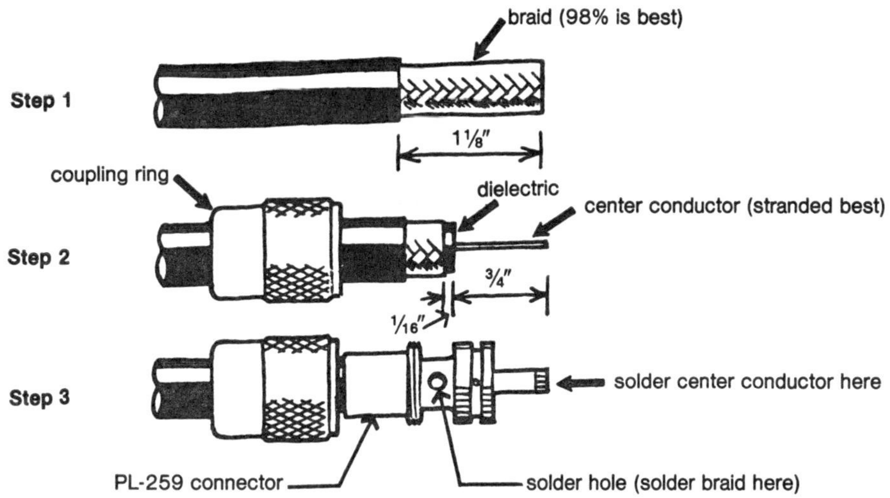

Above: *Proper method of installing a coaxial connector on RG-213/U cable. This cable connector is used with VHF, SSB, Loran-C, and SatNav antenna cables. Proper soldering requires experience. The alternative is to use crimped, no-solder connectors, which can be installed with pliers.* ***Below:*** *Proper method of installing a coaxial connector on RG-8/X cable. This cable is used for short to medium-length runs with VHF, SSB, Loran-C, and SatNav antennas. Attempt this soldering only if you have done it before. Again, crimped connectors are the alternative.*

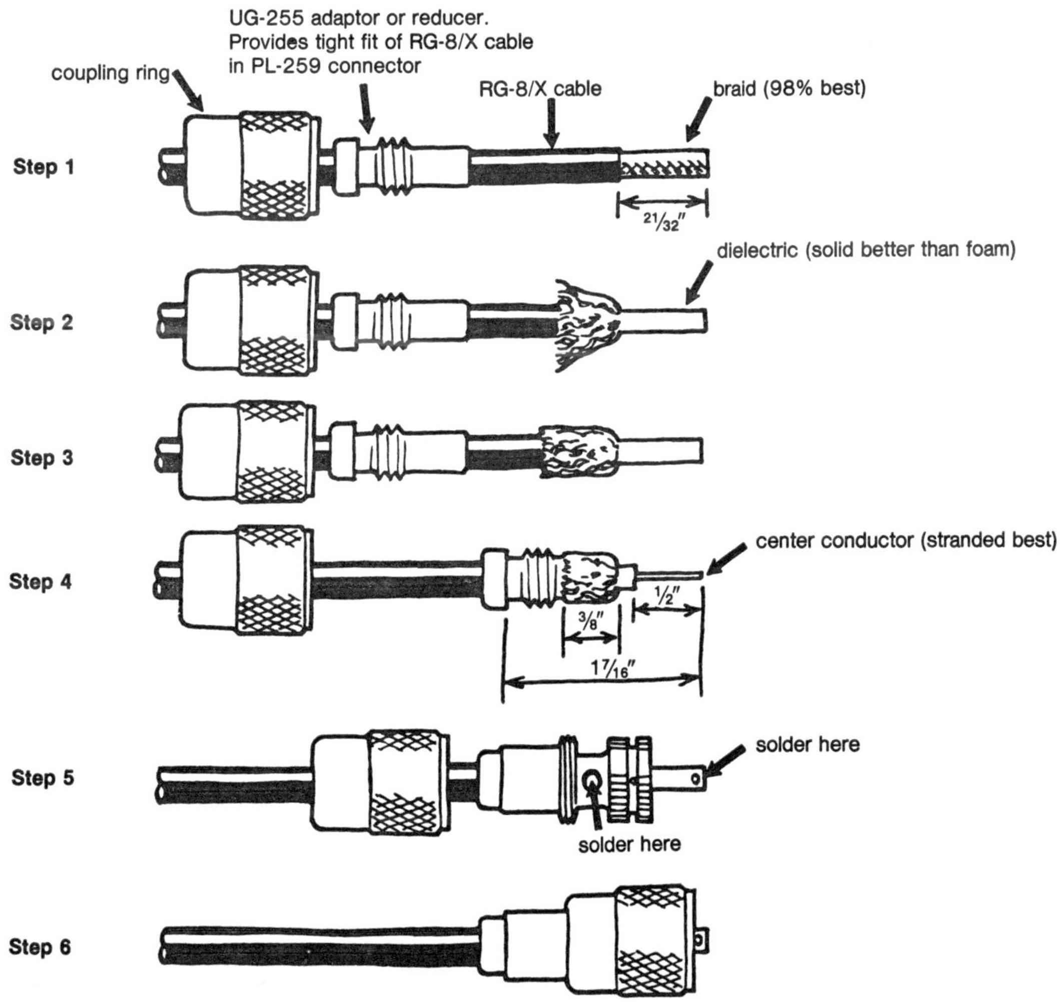

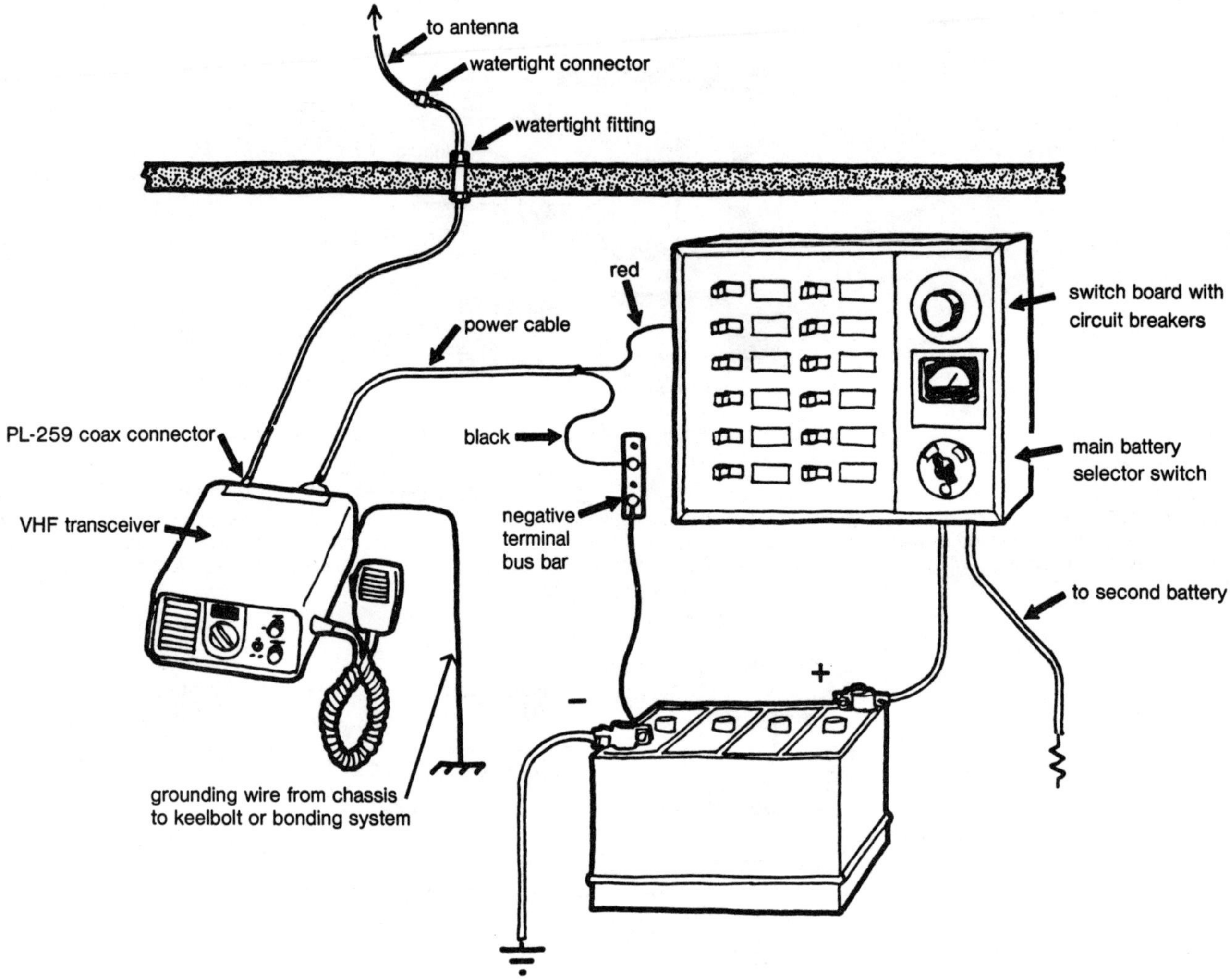

Typical hookup of a VHF radio through a switch panel. An alternative to wiring through an electrical panel is to connect the radio's power cable directly to the battery, making sure fuses are in line with the radio.

Once the cable connector is in place, you will want to test the continuity of the connection. This test cannot be done with regular ohmmeter equipment. Since most VHF antennas look to a DC meter like a direct short, it will take a specialized antenna test meter to doublecheck the operation of your antenna as well as the connection to the PL-259. While some inexpensive CB-type meters for antenna checks may work, you really need an antenna test meter calibrated for VHF frequencies. A simple way to test your antenna connection without a meter is to insert the center pin of your plug in the VHF set socket. Listen for the weak weather station signal. Now push the remaining shell of the connector assembly onto your antenna receptacle, and the weather station should increase slightly in signal strength. If it does, your connection is proper. If you slide the shell onto the connector and the weather station disappears completely, you've got a problem with your connection or with the VHF antenna itself.

In a powerboat, mount the radio where it's easy to see and operate from the helm. Mount it handy to the companionway in a sailboat. Unless it's a waterproof or weatherproof radio, keep it out of any area that might get wet. Keep any external speaker at least 3 feet from the compass. Place the microphone hanger at a convenient location, but one where the mike is not likely to be left switched on.

Run the red and black 12-volt wires supplied with your VHF set to your switch panel. Rather than tacking them onto any red and black wire you might find at your navigation station, go directly to a healthy 12-volt source that feeds voltage selector switches. Use a voltmeter to doublecheck that there is no more than a half-volt drop when you transmit on high power into a dummy load (nonradiating) test antenna network. If the dial lights dim dramatically when you transmit, the voltage input to the set is inadequate.

If you purchased your set with a remote control unit for a flying bridge, run the wires as recommended by the manufacturer. Manufacturers' instruction books are usually excellent, and you should follow their suggestions precisely.

After you have doublechecked all connections, start tuning around the channels and listening for distant stations. Engine noise and fluorescent lights do not affect VHF radio as they do SSB and AM radios, but hot or noisy ignition circuits can produce signals that, although not apparent as background noise, will "piggyback" on weak incoming broadcast signals, confusing them. If weather channel signals transmitted 100 miles away come in loud and clear, everything is fine with your set. Don't transmit until you obtain proper FCC licensing (see below).

HANDHELD UNITS

VHF handhelds are terrific performers. A handheld is not recommended as the only marine VHF set aboard a large boat, but one of these little sets could make a great single radio installation aboard a very small boat. The handhelds are usually all-frequency synthesized. Although some manufacturers still try to peddle crystal-type VHF radios, both stationary and handheld, your better buy is a synthesized set that is just as reliable and picks up every channel.

VHF handhelds may use either keyboard-entry synthesis or thumbwheel channel selection. The thumbwheel channel selection method usually leads to a lower price tag, but the thumbwheels are very leaky and must never get wet. Keyboard entry usually leads to a more weathertight handheld that can survive a little bit of spray. No handheld will take a complete underwater dive.

Some units offer a choice of battery packs. A high-

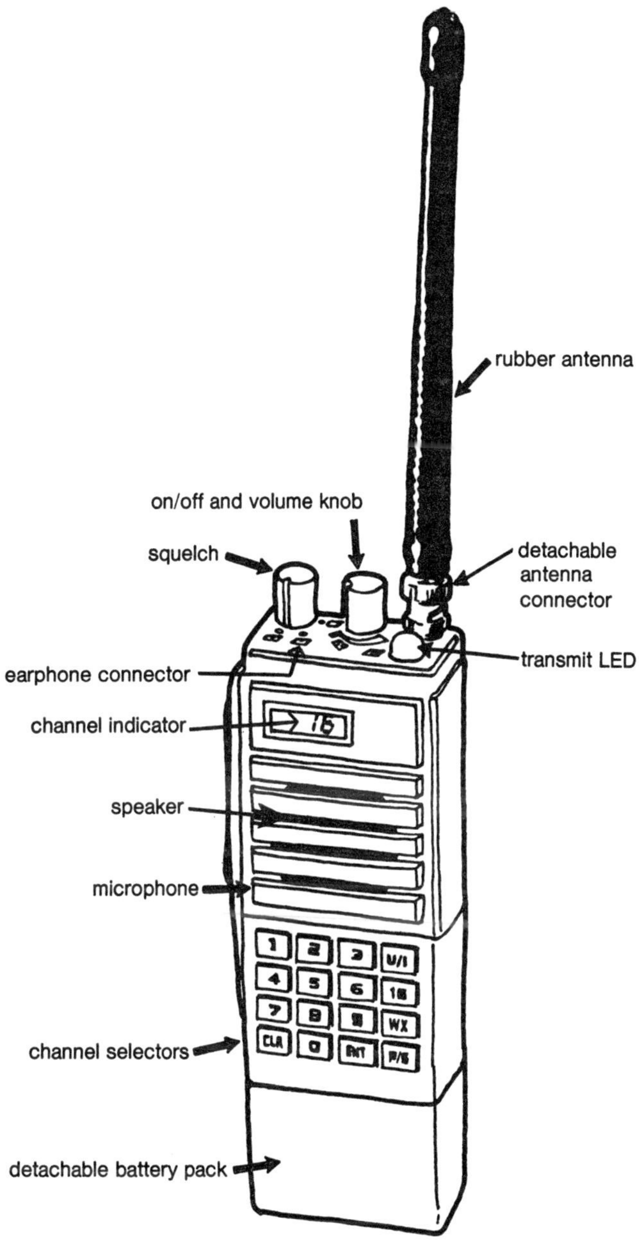

Typical handheld VHF radio with self-contained batteries. Some units are all-channel in their capabilities, while others have a limited number of channels. Range is shorter than it is with radios using a conventional antenna, although some models will take an external antenna for extended range. The price range is $200 to $500, and a typical size is 8½ inches high, 3½ inches wide, and 1½ inches thick.

power pack may give you 5 watts, but it will wear down after an hour or two of broadcast time. Medium-power (1 to 3 watts) and low-power packs will last progressively longer. Most can be recharged.

You can hook up your handheld to an external antenna, improving its range dramatically over what is possible with the little rubber antennas the units come with. Always use an external antenna on a VHF handheld if it is the only radio aboard.

There are a plethora of accessories for any handheld, but choose only those that match your style of operation. As mentioned, you can even get an external power amplifier that takes the 3 watts output and amplifies it to 25 watts, but it's cheaper to buy a complete VHF 25-watt system in the first place.

The licensing for VHF handhelds is identical to a regular 25-watt marine VHF set. Operating the handheld on shore is disallowed by the FCC.

LICENSING

All radio stations aboard a vessel must be licensed by the FCC. You cannot transfer this license from one boat to another or from person to person. Application for a marine VHF license is on Form 506, which is normally packed with every VHF transceiver from the manufacturer. There is no fee. The completed and signed application is sent to the Federal Communications Commission, P.O. Box 1040, Gettysburg, Pennsylvania 17325. It takes about six weeks to receive your call letters. On Form 506-A there is a special provision that allows you to use a temporary identifier as a call sign the second you put Form 506 in the mail to the FCC. Your call sign will consist of numbers and/or letters that comprise your vessel registration number or documentation number.

You and any regular crew members who plan to use the equipment aboard your boat should possess a personal FCC license, called a restricted operating permit. The application for a restricted radiotelepone operating permit is made on FCC Form 753, and it's also free. You must be at least 14 years of age; there is no test other than trying to figure out all the questions asked on the license application.

Form 506 is also used in applying for a handheld license if you plan to use your handheld on many different vessels, but not one in particular, or if you plan to use it in an unregistered small boat.

OPERATING PROCEDURES

Most instruction manuals include a radio logbook. This is a handy place to keep call sign information of stations you may communicate with, or to write down phone numbers in case you need to make a marine radiotelephone call to shore. Try to keep your log updated, and certainly note any distress messages you hear.

Most manufacturers and some marine electronics sellers include a transmitter certification card indicating that your set has been properly tuned and certified for operation. This important information must be kept with your logbook. If you find that your transmitter has not been certified by a technician, you are required by law to obtain this certification. Don't have a tech come all the way down to the boat and charge you an arm and a leg for doublechecking your transmitter frequencies, but do try to obtain a certificate from the original seller or directly from the manufacturer. The days of the FCC walking the docks and checking logs are over, but keep it legal.

Good operating procedures are necessary to keep communications flowing smoothly on the crowded marine VHF channels. Over the last few years, the VHF marine radio channels have become even more overcrowded and even more undisciplined. The only time you are permitted to talk about anything other than relevant ship business is when you are tied into the marine operator. CB jargon is inappropriate and illegal. Random channel-hopping is also illegal. By law you must stick with channels that agree with your type of operation.

It is also illegal to call the Coast Guard for a radio check. They have such sophisticated monitoring equipment that any type of signal will sound loud and clear to them. Call another boat instead.

The United States Coast Guard and the Radio Technical Commission for Marine Services (in cooperation with the Federal Communications Commission) recommend standard procedures for establishing communications on marine VHF and high frequency channels. These procedures are detailed in piloting books and in VHF instruction manuals as well as government literature. Contact your local Coast Guard detachment if you can't locate the procedures elsewhere. These operating procedures should be followed.

MAINTENANCE

Little maintenance is required to keep your marine VHF system perking along. Check all outside antenna connections now and then to make sure you have no bare wires or corroded cable plugs. Many VHF antennas bring the coax directly into their interior, so there is nothing to worry about. Base-loaded whip antennas, however, may require an outside PL-259 connector, and it's best to unscrew this connector periodically and check out what's happening inside. If the cable and connector have turned green, replace the feedline and plug.

Any sort of plug located on the exterior of your vessel should be sealed with a flexible silicone sealant. This stuff will make more of a job of removing the connector for periodic inspection, but it will seal out the weather and corrosive elements that might otherwise attack an outside antenna connection.

Check periodically to ensure that all plugs in the back of the VHF set are snug. With the radio tuned to a transmitting weather station, wiggle each wire by turn and listen for a breakup of reception. If the weather station should begin to "break up" when you wiggle a wire, chances are that wire needs to be replaced or its connection repaired. To check the mike cord, establish communications with another vessel and then wiggle the mike cord as you're talking. If your signal breaks up, you may need a complete new microphone setup. Most mike connections go bad where the cord disappears in the plastic mike or where the cord has worked loose from the chrome chassis connector.

On the water, use your VHF set in the receive mode on Channel 16 often. Not only could this very well save a life, it also helps dry up any moisture that might be inside the VHF cabinet. Transistors don't wear out, so keep your set on whenever you're underway.

ADDITIONAL BUYING TIPS

If you know which VHF radio you want, consider buying it from a mail-order discount house. The installation is not difficult (particularly if you use crimped, no-solder coax connectors), and the radio is easily unplugged and shipped back to the manufacturer for repair. Manufacturers will usually repair items and return them within three weeks or so. In this instance, then, you do not really need the support of a reliable local dealer standing behind the manufacturer's warranty.

4

Marine Single Sideband Radio (and Ham Radio)

by Gordon West

Marine single sideband (SSB) radio enables mariners to communicate ship-to-ship or ship-to-shore over much greater distances than VHF transmissions can cover. The maximum possible communications distance is determined by the interplay of many factors, of which the time of day, the choice of marine SSB frequencies, and the quality of the shipboard installation are generally the most important. In favorable circumstances, communications around the world are possible.

If this sounds interesting, you'll also be interested to know that SSB radiotelephones are now available at substantially lower costs with more features than in the past. Marine SSB equipment is more expensive than VHF radiotelephone gear, however, and it is not a replacement for VHF. The FCC requires you to have a permanently installed VHF set aboard before you can add any other marine radio, and your SSB set should never be used when the VHF radio would suffice. To do so would only cause needless crowding on the marine SSB frequencies.

A complete marine single sideband station consists of a transceiver, an antenna, and—where channels on more than one of the seven designated marine SSB bands are to be used—an antenna coupler (also known as a tuner) between the transceiver and the antenna. Both the transceiver and the coupler must be electrically bonded to a proper ground, "proper" in this context implying elaborate and, perhaps, difficult to install. Without a good ground, the SSB station simply will not function adequately. When it works well, however, a marine SSB radio is a provider of entertainment, information, and communications for fishermen, commercial vessels, offshore sailors, and bluewater voyagers; most important, it could be a lifeline in an emergency situation.

HOW IT WORKS

The term "single sideband" designates a type of radio transmission that the military and ham radio operators have been using for years to transmit messages throughout the world. In 1971 the FCC phased in single sideband transmissions for the long distance marine radio service. At the same time they introduced the expanded marine VHF service for local communications and phased out the amplitude modulated (AM) double-sideband sets. A single sideband signal concentrates your voice into a tightly compacted

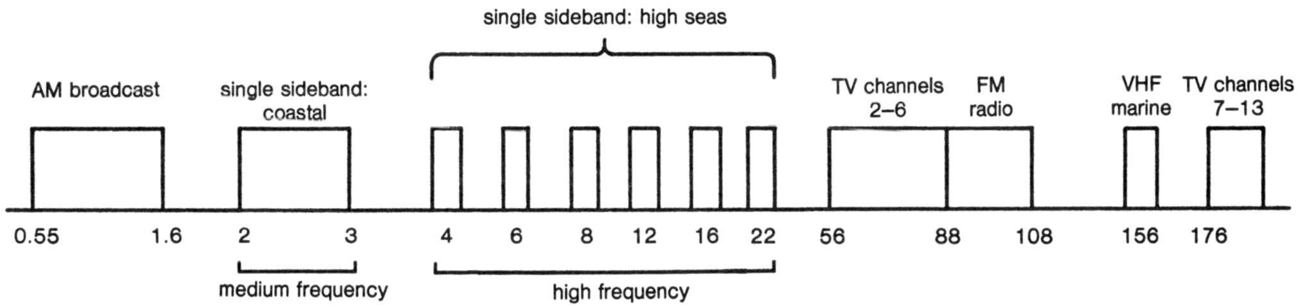

Frequency allocations at a glance.

radiowave capable of traveling hundreds or even thousands of miles. This very efficient radio signal is a faithful reproduction of your actual voice pattern. Unlike a commercial AM broadcast station that sends out duplicate voice waveforms plus an energy robbing "carrier," marine single sideband eliminates the unneeded mirrorlike lower sideband and the power-robbing "carrier" that has no function except to hush background noise when nothing is on the air. All the radio energy from your voice is concentrated in an upper sideband waveform that gives you worldwide talk power. Only when you speak will radio energy jump out into the airwaves, and between each word your transmitter and battery system relax. This system imposes a very low current demand on your storage battery system.

Because the transmitted signal is compacted into a very narrow band width, distant receivers are able to reject almost half the normal noise level from the air waves. FCC-required frequency tolerances keep single sideband sets precisely on frequency to minimize that "Mickey Mouse" effect on receive. Simply by adjusting a clarifier knob on your transceiver, you can produce a normal-sounding voice from a transmission by a distant ship or shore station.

A marine single sideband transceiver broadcasts on one of seven allocated frequency bands: 2MHz, 4MHz, 6 MHz (used mostly by stations on the Mississippi River), 8 MHz, 12 MHz, 16 MHz, and 22 MHz. The 2 MHz band is known as the medium frequency or "coastal communications" band, because frequencies in this band (between 2 and 3 MHz) are typically used for coastwise contacts just beyond VHF range. These frequencies are referred to by their actual values in kilohertz, but frequencies within the high-frequency 4 MHz through 22 MHz bands—referred to collectively as the "high seas" frequencies—are given channel designations.

When a transmission is made on any of the bands, one component of the radio signal hugs the surface of the ocean. This is called the groundwave, and it will travel approximately 100 to 200 miles. The groundwave usually does the work in single sideband communications between a boat and a nearby shore station or another boat less than 100 miles away; it is the functional component of transmissions in the 2 MHz band. Groundwave propagation is consistent day or night, depending for the most part only on a good, strong transmitted signal, though it can be interrupted by thunderstorms or other atmospheric disturbances.

The "skywave" component of the transmitted radio signal enables the long-distance communications for which single sideband radio is noted. Skywaves travel upward, bounce off the ionosphere (a multilayered envelope of ionized gas surrounding the earth), and are reflected back to earth hundreds and sometimes thousands of miles away. The height, ion density, and reflecting capabilities of the ionosphere are dependent on radiation from the sun and thus change with day and night, with the seasons, and with the 11-year solar cycle; the distance covered by skywaves of a given frequency when they reflect from the ionosphere varies accordingly. In general, lower frequencies bounce back to earth close in, while frequencies in the 12 MHz band reflect over fairly long distances, typically 3,000 miles. In the 22 MHz band you may be able to communicate from the West Coast of the United States into the Mediterranean. If the ionosphere is strong enough, the skywave will often bounce a second time, doubling the maximum possible communications distance. You can easily talk all the way around the world on 22 MHz at such times.

During daylight hours the ionosphere rises and its ion density increases; the range available on the higher frequencies increases accordingly. The skywave

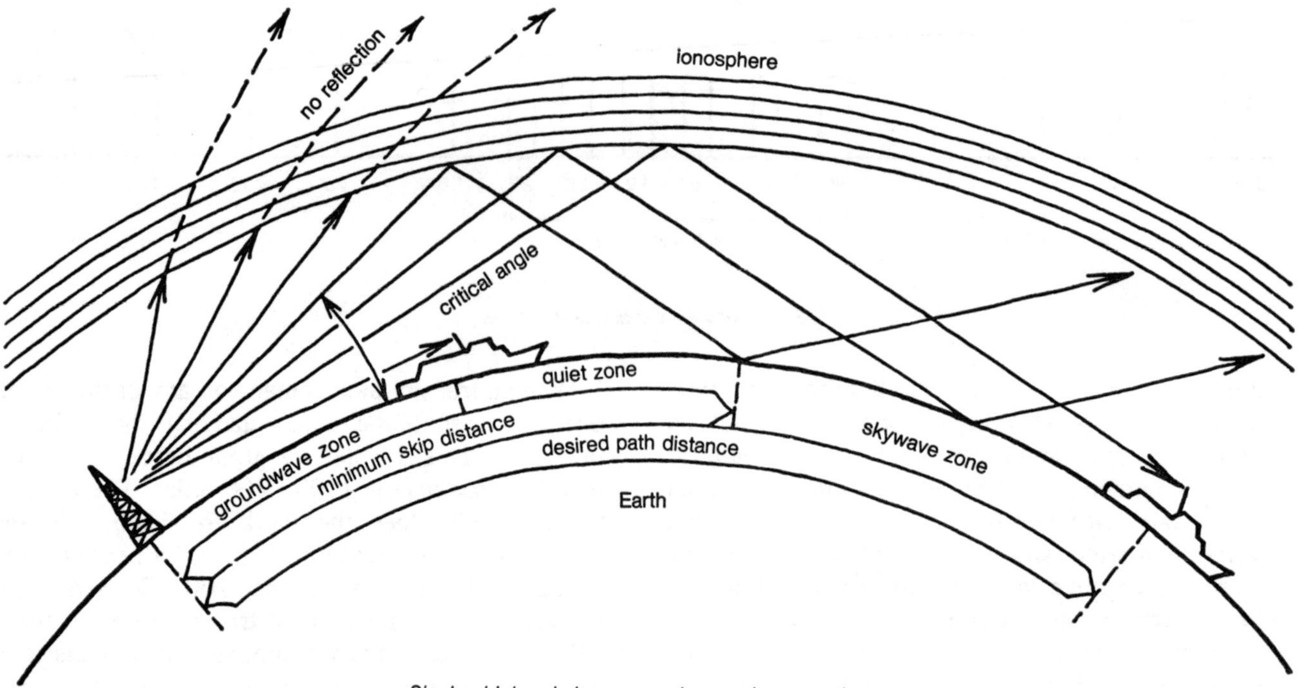

Single sideband skywave and groundwave paths.

components of lower-frequency transmissions are absorbed by the ionosphere, however, while their groundwave components are inhibited by the higher daytime noise level near the earth's surface. Maximum ranges of 2 MHz, 4 MHz, and 8 MHz transmissions tend to decrease during the day.

During the night, the ionosphere gradually lowers and its ion density decreases. Skywaves in the 16 and 22 MHz bands may not be reflected at all after the evening hours. Nonetheless, you may still be able to enjoy a communications range of several thousand miles using one of the lower frequency bands.

Skywaves are unaffected by local weather conditions.

After a few weeks of playing around with a single sideband radiotelephone, a user will begin to get a feel for the expected range on any one particular band of frequencies. As a rough guide, here is what to expect for ship-to-ship communications in the groundwave range: 2 MHz—150 miles (perhaps more at night); 4 MHz—100 miles; 6 MHz—75 miles; 8 MHz—70 miles; 12 MHz—50 miles, and 16 MHz—50 miles. For comparison, a VHF radio (156 MHz) might give you about 40 miles ship-to-ship. The following table will give you an idea of what to expect from the skywave component of your SSB transmission.

Frequency Band	Daytime Range	Nighttime Range
2 MHz	Skywaves absorbed	1,000 miles
4 MHz	Skywaves absorbed	1,500 miles
6 MHz	500 miles	2,000 miles
8 MHz	700 miles	3,000 miles
12 MHz	1,500 miles	Worldwide in the direction of the sun.
16 MHz	3,000 miles	Worldwide in the direction of the sun until 8 P.M. local time, becoming inoperative after that.
22 MHz	Worldwide	Little skywave reflection after sunset.

To increase your maximum communication range, you should go to a higher frequency. If your signal is literally skipping over the desired station, however, you should switch to a lower frequency. In general, the greater the maximum communication range afforded by a particular band at a particular time of day, the broader the intervening skip zone.

SSB SERVICES AND BROADCASTS

For any mariner outside VHF range, the marine single sideband service can be a lifeline in an emergency. The United States Coast Guard and other distress agencies throughout the world monitor 2,182 kHz (in the 2 MHz band) as the international distress frequency, allowing you to contact shoreside and marine rescue agencies immediately when outside VHF Channel 16 range. There are literally thousands of stations guarding this channel for a distress call 24 hours a day. The U.S. Coast Guard includes five additional working channels—424, 601, 816, 1205, and 1625—in their Automated Mutual-Assistance Vessel Reserve program. Imagine using your marine SSB set to place a call for help when you're thousands of miles away from any shore station. Through their AMVER system, the Coast Guard can pinpoint the position of commercial and military vessels passing through your area and signal them to change course and steam to your location to render assistance.

A marine SSB set gives access to other services besides emergency assistance. Shoreside commercial telephone stations stand by on dozens of frequencies, ready to place a mariner's phone call. These shore-based phone companies operate elaborate transmitting and receiving antenna systems to enable clear communications, and they transmit "traffic lists" to ships at sea that have telephone calls waiting from shoreside parties. They also broadcast weather reports, storm warnings, and other safety-related notices to mariners, and with their massive antenna systems they can patch a mariner into rescue coordination centers, hospitals, and emergency-at-sea medical systems, without charge, should an emergency arise.

Incidentally, shoreside marine telephone stations also make it easy to predict the best band to use for rock-solid communications. Every four hours when they read the traffic lists and give the ocean weather conditions, they transmit simultaneously on each one of their authorized bands. By switching bands while they are transmitting, you can determine which band offers the best reception. Where you hear them best is where they will hear you best.

Private shore stations share the ship-to-ship channels, meaning that a mariner can communicate directly, at no charge, with, say, a marine supply company. Other "limited coast stations" might include your own marine business, yacht clubs, marine salvage companies, and private air ambulance companies—in short, any marine business that needs to communicate regularly with distant ship stations. You can even establish a base station at your office to stay in touch regarding marine matters when you're far out at sea.

Depending on the frequency capability of your SSB transceiver, you may be able to receive other services that share frequencies adjacent to the marine band channels. For example, it is possible to tune into worldwide international broadcast stations and find out the latest news here and abroad. You can eavesdrop on the military, State Department, and foreign embassy communications that fill the worldwide high-frequency spectrum. You can tie a weather facsimile receiver into a marine SSB set and receive crystal-clear weather charts of your cruising area. You can tune in amateur radio frequencies and hear local weather reports through one of the many maritime amateur radio "nets." (If you hold a valid ham license, general class or above, you can also use a marine single sideband transceiver with broad frequency capability to transmit on any amateur radio frequency.) And you can pick up the international time signals wherever you cruise, using one of many different frequencies. Tick, tick, tick, at the sound of the tone, it is exactly . . . . You may also be able to receive transmissions originating from U.S. Air Force flights, the Strategic Air Command, Air Force 1 (the president's plane), Civil Air Patrol, United States Intelligence Agency, Antarctic stations, Interpol, U.S. weather ships, the Hurricane Research Center, and Volmet-Aviation weather broadcasts. A single sideband marine transceiver with all-frequency receive capabilities will literally tune in the world.

SSB EQUIPMENT

Transceivers

In sharp contrast with most other marine electronics, the Americans have dominated the manufacturing and marketing of single sideband equipment. While a handful of single sideband manufacturers have created some excellent marine single sideband equipment, big leaps in technology and quantum jumps in price reductions have been few and far between. The switch from crystal-type to synthesized, no-crystal single sideband sets has taken place very slowly among

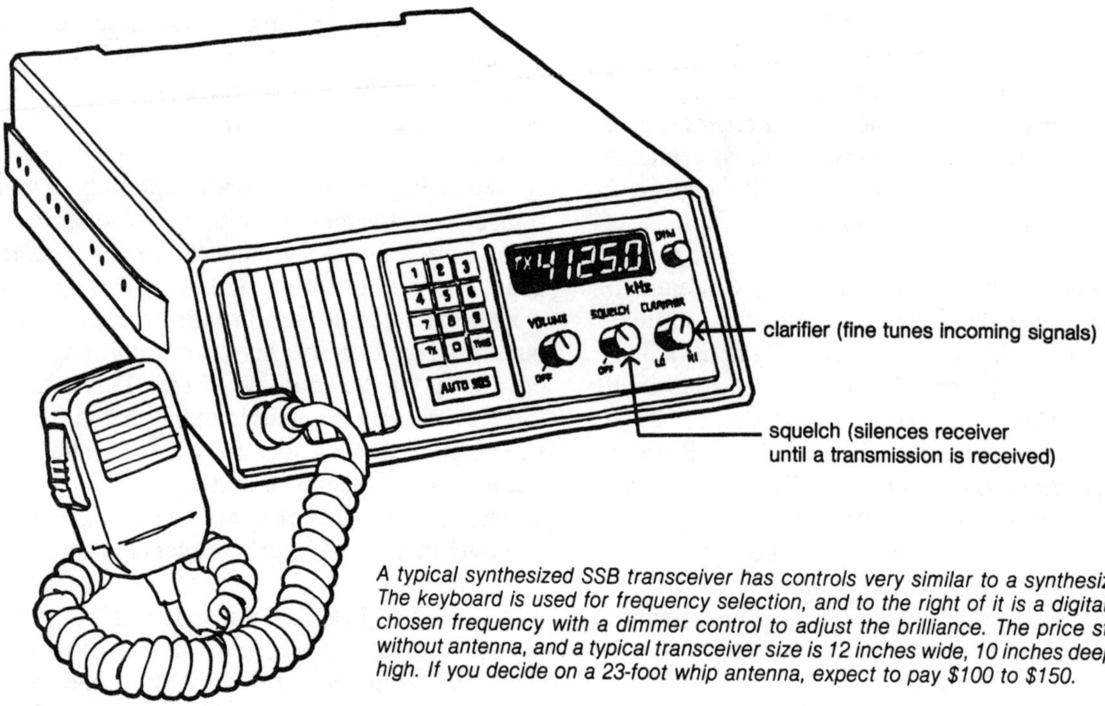

A typical synthesized SSB transceiver has controls very similar to a synthesized VHF radio. The keyboard is used for frequency selection, and to the right of it is a digital display of the chosen frequency with a dimmer control to adjust the brilliance. The price starts at $1,500 without antenna, and a typical transceiver size is 12 inches wide, 10 inches deep, and 5 inches high. If you decide on a 23-foot whip antenna, expect to pay $100 to $150.

American manufacturers, and the prices of both types have been high. Most sideband stations, complete with tuner, have been selling for well over $5,000.

Recently, however, the Japanese have entered the manufacturing picture, and in just one year they completely revolutionized the thinking on marine single sideband among manufacturers as well as purchasers. The very first Japanese-manufactured single sideband set, type accepted by the Federal Communications Commission, featured 48 programmable channels, 125 watts of peak power output, capabilities for any marine channel between 2 MHz and 22 MHz, and a price tag under $1,500 not including an automatic antenna tuner and some sort of a whip or long wire antenna system. With the latter items included, the complete single sideband station still sells at half the price American manufacturers had been charging for sets with half the features.

Within 60 days after the debut of the 48-channel, completely synthesized, $1,500 marine single sideband set, American manufacturers and their dealer system dropped the price of their keyboard-entry synthesized SSB sets by $1,000. No doubt American manufacturers are taking a hard look at this new Japanese SSB set that's more powerful, has more channels, is priced more competitively, and is an equal performer to what's made in the U.S.A.

The price of a synthesized set with broad frequency capability is unlikely to go much lower than $1,500, but we can expect to see more Japanese-style transceivers hitting the U.S. market. Indeed, two other sets with similar capabilities for similar prices appeared in early 1985. There is absolutely nothing wrong with Japan's marine electronics technology. They simply have a lower paid labor force that can turn out a ton of product at an extremely low cost to a manufacturer-importer in the States. When production overtakes distribution, prices drop, and we get the benefit of high technology equipment at low prices.

In light of recent developments, a synthesized SSB set has to be considered a better buy than a crystal set. With the former it's easy to program frequencies and channels. Instead of purchasing expensive additional crystal elements—one for transmit and one for receive, a pair for every channel you want—and paying an FCC-licensed technician to install them, you can let frequency synthesis do all the work with just one reference crystal. The crystal is built into the radio, and you have only to keyboard enter the frequencies and channels you may wish to tune. Most mariners need about five frequencies in each band, and an overall total of 30 memorized frequencies is usually plenty. The older-style crystal radios offering 11 channels simply don't offer enough frequency capability to satisfy the needs of most cruising mariners.

A synthesized set that will "memorize" more than 30 channels will allow you to preprogram not only marine frequencies, but also some of those receive-only worldwide general coverage frequencies mentioned earlier. Licensed ham operators can even program transmit-and-receive marine net frequencies. Old channels can be "erased" in favor of new ones at will.

Most single sideband transceivers offer at least 100 watts output, although there are some that put out 20 watts or less. Higher output means a stronger, more solid signal provided the shipboard installation is a good one.

Antennas

After a good groundplane, the antenna is the second most important requirement for a solid transmitted signal. For powerboats, a fiberglass whip antenna is usually the best choice and easiest to install. Two models are available—a 23-foot "trap" antenna that doesn't need a tuner, or a 23-foot white fiberglass antenna, without traps, that does need a tuner. Either works well when tied into a good ground system, but most mariners choose the 23-foot white fiberglass whip without traps, and mount an antenna tuner below the antenna base. The trap antenna has high efficiency, but it is also fairly expensive—about $1,000—and usually needs to be "matched" after the installation, a job which must be performed by a technician.

Sailboat owners frequently elect to use a backstay or a hoisted wire (No. 14 plastic-covered copper wire is good) for an antenna. Either alternative is inexpensive, avoids the wind resistance and what some would consider the unsightliness of a long whip, and frequently gives better results. Insulators must be placed in the bottom and top of the backstay, but these incur no loss of rigging strength. If a dismasting occurs, however, the antenna will be lost when it is likely to be most in demand. A spare whip antenna should be carried for such a contingency.

An automatic antenna coupler is almost always necessary to tune up an insulated backstay or hoisted wire. With a backstay antenna, the coupler is mounted aft in the lazarette; a foil grounding strip goes to one side of the tuner, and the "hot wire" goes to the backstay on the other side. If you choose the microprocessor-controlled, relay-switching tuner, you will also need a tiny current draw from a 12-volt source to work the relays on the tuner.

There is no magic length of random wire for any particular frequency. The tuner will electrically lengthen or shorten a backstay or hoisted wire. Keeping the antenna in the clear is a big consideration, however. If your backstay or hoisted wire antenna is surrounded by a lot of other stays, you are likely to lose some energy into your rigging. By the same token, when a powerboat whip is mounted too close to a grounded tuna tower, valuable energy is lost.

Antenna Tuners

Unless a pretuned, prematched, 23-foot-whip trap antenna is used, any marine SSB set will need a separate antenna tuner to match the impedance of a whip or backstay antenna system to the set for frequencies between 8 and 22 MHz. (For the 2 and 4 MHz frequencies, most single sideband transceivers have built-in circuitry to couple to a 23-foot fiberglass whip or to a backstay.) The usual antenna tuner or coupler is a solid-state instrument that is weatherproof and mounts out of the way beneath the antenna system, and it can range in price from $600 to $1,500. Fair results are obtained from a $600 resistive coupler, but the $1,000-to-$1,500, relay-switching, microprocessor-controlled couplers will achieve the best possible match to squeeze the last bit of power out of the system and into your antenna setup. Top-of-the-line couplers have almost unlimited frequency capabilities.

There are also manual antenna tuners, which are available for as little as $100, but these necessitate your having to fiddle with three knobs every time you change bands. It takes skill to keep these tuners matched to the transceiver, and they are usually of interest only to amateur radio operators. (They are available only in ham radio outlets.) If you do decide on one of these, any ham radio-type 50-ohm antenna tuner will work well providing the coaxial cable feedline antenna system offers a feedpoint impedance of 50 ohms. Your set will work just fine with a manual tuner.

GROUNDING (AND BUYING TIPS)

Good grounding is a must for good single sideband

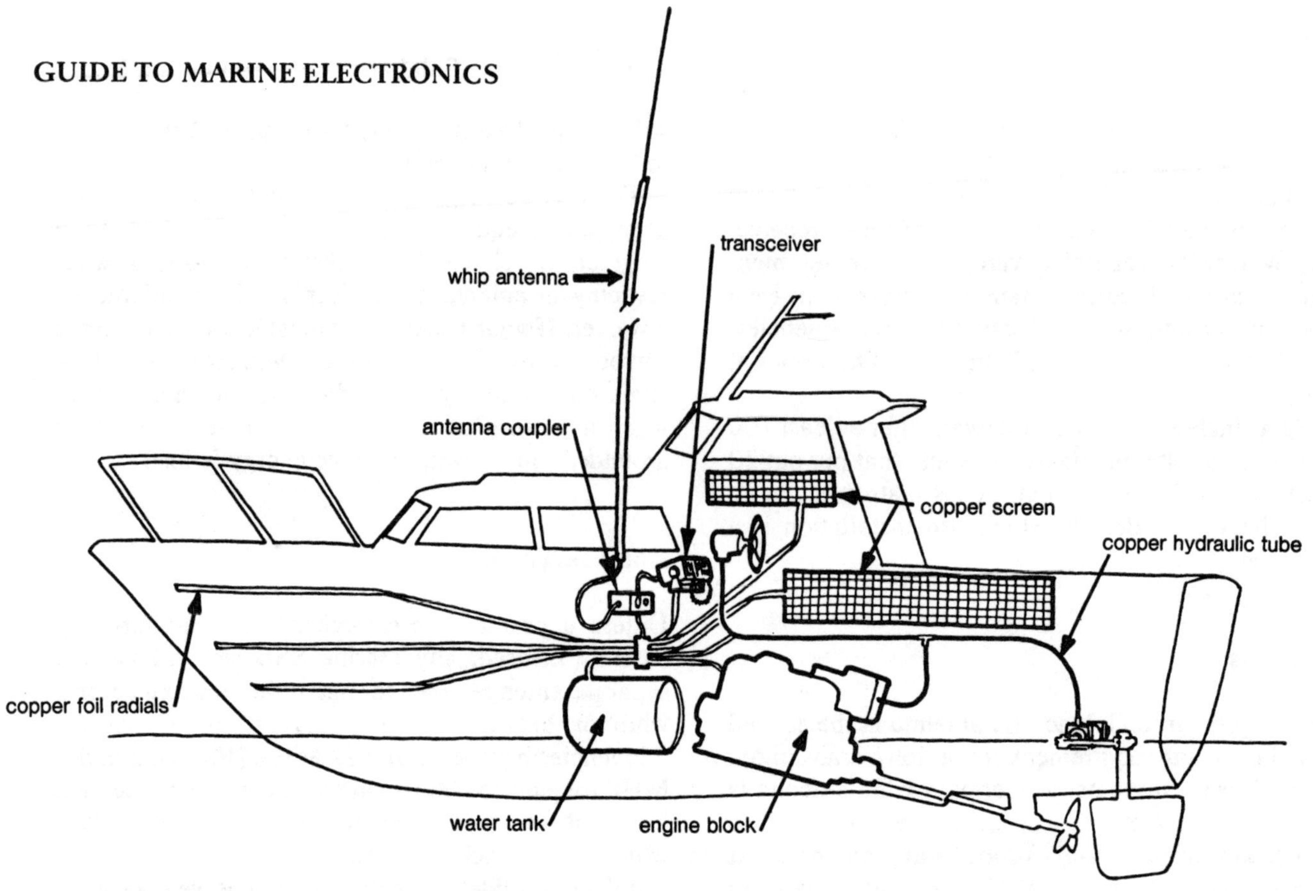

Possible components of a powerboat SSB groundplane system. Good grounding increases transmission and reception range. Copper foil works better than copper strap for radial strips and for bonding of groundplane elements.

range, since half of your antenna system is your RF ground. Your antenna is merely the radiating portion of the entire antenna system, and it needs to see a mirror image of itself before it will send out an SSB signal. This mirror image, called a counterpoise, is created by using the seawater as a groundplane. Installing an adequate grounding system is not simple, and you may wish to have an FCC-licensed technician do the work. If you do, expect to pay your local marine electronics dealer *at least* $700 for the work (more for a more complicated installation) *unless* you buy the transceiver from that dealer and include the installation as part of the sales package. This removes the installation charge from a strict time-and-materials basis, and it's certainly a wiser course than buying the radio from a discounter and then hiring a dealer for the installation. Insist on having the installation agreement in writing, however—along with the servicing warranty terms—and insist on further work at no extra charge if the initial installation is unsatisfactory.

A compromise solution is to buy from a dealer, do the time-consuming part of the installation yourself (running copper foil and screen throughout the boat, and grounding to keel bolts, etc.), then get the dealer's technician to solder the connectors, install the transceiver, and check out and certify the installation. You'll save some money, but you'll probably also lose the guarantee of further work at no charge should the installation prove unsatisfactory. In any case, what follows is a description of the grounding system.

The groundplane system begins directly below the antenna feedpoint. (When an antenna coupler is used, it should be considered the "feedpoint.") An ideal ground for all-frequency single sideband work consists of 60 to 100 square feet of metal surface area. If this sounds like an impossible area to achieve, consider the following possible components: an encapsulated lead keel; metal water tanks; stainless steel tuna towers and stanchions; chainplates; propeller and shaft; through hulls; external grounding plates; and the engine block. In a fiberglass or wooden boat, as many of these as practicable will ordinarily be incorporated in the grounding system. A more radical alternative is to fiberglass into your hull large amounts of copper screen and 3-inch-wide copper foil strips (available through most marine electronics dealers). Of course,

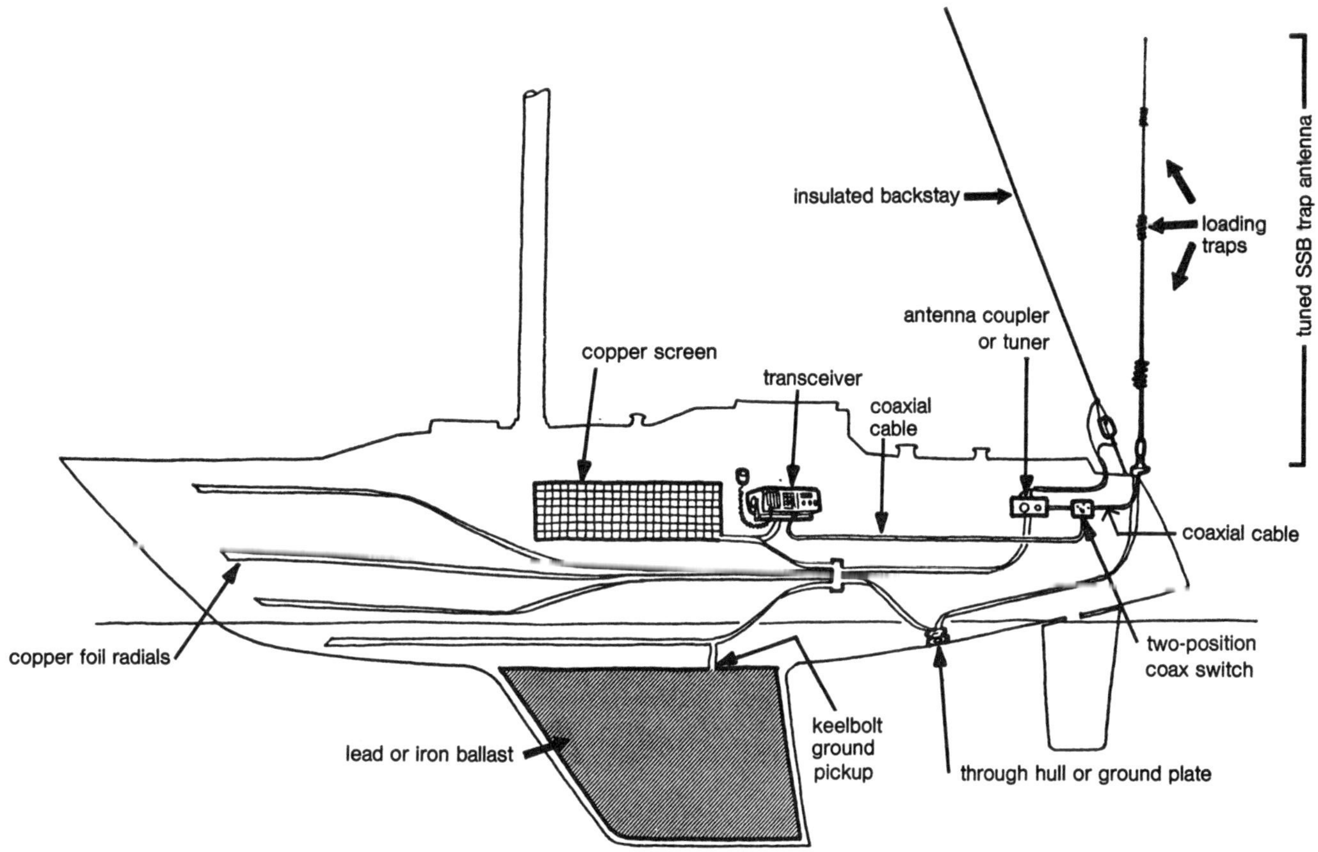

Possible sailboat groundplane system. The best and largest ground component when available is the lead or iron keel, which, even if internal, provides a capacitive ground. Two alternative antenna systems may be required in racing sailboats by the rules governing offshore events. The trap antenna shown here would be installed if a dismasting were to occur.

the hulls of aluminum or steel vessels provide unsurpassed groundplanes.

Use copper foil to connect your transceiver and antenna coupler to the grounding system. Round wires are almost "invisible" at radio frequencies, so battery cables or other small wires for grounding won't work. Where the copper contacts dissimilar metals, an intervening stainless steel washer will reduce the potential for electrolysis. The joint can then be painted over to seal out water. Your ground system does not actually need to contact the seawater to be effective. An encapsulated lead keel, for example, makes a "capacitive" ground by virtue of its proximity to the seawater. Double bolt the copper foil to an exposed keel bolt or tap directly into the keel with a bolt going through the copper foil and into the lead.

Use stainless steel clamps if you wish to attach foil to through hulls, and remember to run foil between every through hull. Hose clamps will also make a good, tight connection to water tanks and copper hydraulic lines.

Grounding to underwater metals that are already tied in with a common groundwire will not affect the existing corrosion control system. If your boat does not have such a system, you may wish to avoid grounding to engine blocks, through hulls, and other metals in direct contact with seawater.

If your antenna and tuner are mounted high atop the bridge, and you must create your own grounding system, remember that it must begin directly at the tuner. This may require running copper foil and copper screen inside the headliner of your pilot house.

When you think you have approximately 60 to 100 square feet of ground below or near the waterline, one foil strip is run to the single sideband transceiver and attached under the ground lug on the chassis. A second copper foil lead goes directly to the antenna tuner. The tuner will usually have a ground terminal, to which the foil is attached.

Achieving a good ground system is a bit difficult, especially if you can't get at your keel bolts. In such a case, consider drilling into the keel. Any sailboat system that doesn't use a poured keel is losing a tremendous advantage in potential signal strength. Only if your keel is made of lead shot poured in fiberglass should you elect not to use it. In any other case, a keel will save you many hours of trying to run copper foil and screen belowdecks.

Good RF grounding techniques will also enhance your overall protection from a lightning strike. The two have a common denominator—a large amount of conductive surface area below the waterline.

Again, small wires—even battery cables—are not effective for RF grounding. Use foil or strapping, preferably foil. Aluminum foil will work in a pinch, and you can even use aluminum air-conditioning foil with stickum on the back.

TRANSCEIVER INSTALLATION

The very compact SSB transceiver occupies little room. Everything is contained within the chassis, which can nestle right alongside the other nav gear in a typical station. The sets are not waterproof, so keep them clear of portholes or hatches that might admit water. You can, however, build the equipment into a wheelhouse instrument panel without worry of lack of ventilation. Since SSB radios are transistorized, the slight amount of heat they develop merely dries out their insides. You may well want to keep the equipment down low for easy channel selection, making it comfortable to operate. Then, some night in a cozy harbor, you can simply flip through the worldwide frequencies to pick up some action. A heavy-duty mounting bracket is shipped with each rig.

Each transceiver is shipped with a red and black power cord. This is the 12-volt connection, and it is fused in case you should get turned around on polarity. A 150-watt marine single sideband transceiver will draw approximately 18 amps on voice peaks. Only when you talk is current consumed in these proportions—so don't worry, you're not drawing a continuous 18 amps out of your battery while the mike button is pressed down!

It's recommended to hook up your 12-volt connections directly to your ship's battery system. This allows you to stay on the air in case of a malfunction of your electrical panel, which is when you may need your set the most. If you have some hefty 12-volt wires leading from your buried battery compartment to your fuse panel, however, you may prefer to go ahead and make your connection at the panel. Clip off any large amount of excess power cable, but leave just enough coiled up behind the radio so it can be pulled from its mount for adjustments or examinations while turned on.

Route your power cable along the same track as your ground foil, taking care that the sharp foil edges don't nick the cable. Don't use the ground foil as the black side of the power cable—these are two separate "ground" systems. Routing them together creates a neat installation, and the foil partially shields the power cable from ignition interference.

Use wire lugs to attach the cable to a terminal strip on an electrical panel or, if the power cable does not already enter the console, on the back of the radio itself. Since the radio power lead is already fused, you do not necessarily need to go through an external circuit breaker. You can if you want, but that adds one more "weak link" in your power cable assembly.

If you run the power cable to your battery system, choose a battery that is less apt to fail in an emergency. A battery located above the waterline, for example, is less vulnerable to flooding.

If you need to extend the wires supplied by the factory, use No. 6 insulated two-conductor, and make certain that any splices are well-soldered or perfectly crimped and are well protected from the salt environment. Crimped (versus soldered) connectors are discussed in the chapter on VHF radios.

ELIMINATING NOISE

Once your SSB station is installed, turn it on and start listening to the bands. There will be more atmospheric noise on the lower frequencies than the higher frequencies. With your engines and other motors all turned off, this noise is just the usual background static present on every band until a signal appears. Strong signals will usually completely mask the noise, but weak signals on 2 and 4 MHz will only quiet it by about 50 percent. The more sensitive your receiver, the more atmospheric noise you are going to pick up. This is normal; any attempt to filter out atmospheric noise would cause distant radio signals to fade away, too.

The type of noise that can be eliminated is that generated by your engine ignition system or by other onboard motors. Fluorescent lights also create noise that is usually heard on the lower frequencies. Onboard noise sources can be filtered at the spot where they are generated. There are filters for alternators and filters for fluorescent lights. You can put resistor spark plugs on your gas engine and tachometer filters on your electronic tachs. Fuel pumps and bait tanks can be silenced with appropriate filters.

Tune in a relatively weak signal on your SSB set, and then start up the engine. If the signal is still there, your noise problems are few. If the signal disappears—especially if it's a strong signal—you will need to identify noisemakers and get a filter for each. For a complete list of noise-elimination filters, see a marine electronics specialty dealer

When noises external to your boat, such as a passing skiff with an outboard, can be heard clearly on your SSB set, simply depress the noise-blanker button on the front of the radio. This will cancel out the repetitious popping sound almost completely. While it may also help dampen repetition-type noise originating on your boat, noise filters at the source are a better way to go. As in plugging leaks, you must methodically get every single one.

GOING ON THE AIR

You are required by the FCC to have a licensed marine VHF set aboard your vessel before adding marine single sideband. The same call letters are used for both, but you may need to modify your license. Dig out your marine station license and look at the authorized frequency privileges. In addition to 156-158 MHz, does it include 1,800-4,000 kHz and 4,000-23,000 kHz? If not, you will need to amend your present station license using FCC Form 506, which is discussed in the chapter on VHF radio.

Once you have mailed this form into the FCC, you may use your presently assigned FCC call letters over the air and note in your logbook that you applied for additional frequencies.

Your personal operator's license (the one you obtained for VHF transmissions by filling out FCC Form 753), covers you for SSB frequencies, too. Since this restricted operator's permit is good for life, no further action is necessary.

Assuming you have a synthesized transceiver, it will have been preprogrammed by the dealer or distributor that sold you the equipment. Very likely the preprogrammed channels will closely follow the recommended channelization for a particular cruising area. Reprogramming different frequencies is easy; simply refer to your owner's manual for programming instructions.

If you followed precisely the installation instructions for transceiver and antenna tuner, you probably won't need a technician to come aboard and test everything out. The instruction manual that came with your SSB lists several ways to verify full power output. Unless the manufacturer included FCC certification with your unit, however, you *do* need a technician to "sign off" in your logbook and provide the FCC certificate.

Before transmitting on any frequency, listen! In fact, spend a complete week listening to different frequencies in different bands to get a feel for the protocol of marine single sideband communications.

When listening to ship-to-ship and ship-to-private shore station calls, you generally will hear both sides of the conversation. When tuning into the ship-to-shore marine telephone station, however, you will only hear the shore station side of the conversation. This is because ship stations transmit on a separate (duplex) frequency when speaking to marine operators. When your turn to talk to a shore station telephone company arrives, the very professional marine telephone operators and their service technicians will ask you the necessary questions about where you are, who you are, and what number you want. Follow their instructions and you will have no problems communicating through the telephone service. The same holds true for emergency communications with the U.S. Coast Guard AMVER stations.

Your first call will probably be for a radio check. Since it's illegal to call the United States Coast Guard for this purpose, try the distant high seas marine operator. Wait until they are finished with their local weather reports before giving them a call, choosing the band that sounds the strongest to you. Marine telephone companies, if they're not really busy, are more than happy to accommodate a radio check.

You can also receive radio checks from pleasure boats that you might hear on ship-to-ship frequencies. If a station sounds weak to you, they will probably say that you are weak to them. The same applies to the telephone service—if they're not coming in strong, you won't be either. Most commercial vessels will probably ignore any calls for a radio check.

Weak signals are not necessarily a result of

something wrong with your installation. Sometimes ionospheric conditions simply won't favor any band. Try the next band up—and the next band down—to improve signal reports. Try a different time of day, and expect that some days you'll have better signal levels than others. Since SSB radiowaves are dependent on ionospheric conditions, it's quite normal for signal levels to change. You may also notice that signals will fade in and out on the higher frequencies—those in the 12, 16, and 22 MHz bands. Again, this is completely normal and should result in almost no loss of intelligibility during a call.

Another way to check the high-frequency operation of your equipment is to try and receive as many foreign broadcast stations as possible. These stations should come in loud and clear at the proper times but are subject to brief ionospheric fades. If you are hearing plenty of activity on these frequencies, plus strong signals from other boats and shore stations, your installation is probably fine, and you will enjoy worldwide communications with single sideband equipment.

If you do decide to have an FCC-licensed technician check out as well as certify your equipment, a local marine electronics dealership will likely be happy to send down a tech with the proper field-strength equipment to "sign off" your station (at a time-and-materials cost). If you installed the equipment yourself, the technician will check out your antenna tuner setup, doublecheck all connections to ensure that they are weatherproof, make some field-strength measurements, and exchange signal reports with distant stations. Since an electronic technician is quite familiar with the characteristics of single sideband frequencies, he'll know when your set is on the air and operating perfectly. If there is any way to squeeze a few more watts out of your system, he will do that, too.

There are approved procedures for making both normal and emergency calls on the marine SSB radio, procedures developed by the Radio Technical Commission for Marine Services in cooperation with the FCC. These are set forth in the Marine Radiotelephone Users Handbook, copies of which can be purchased for $7.95 from the RTCM, P.O. Box 19087, Washington, D.C. 20036. Your local Coast Guard station and books about piloting and seamanship are other sources of this information.

Before placing telephone calls, you will need to register your station with the appropriate marine telephone operators. If you plan to cruise extensively throughout the world, it's best to register with all four of the U.S. coast stations providing high-frequency "high seas" service. This enables you to make a phone call and have them automatically bill it. The four stations and their addresses are: Station KMI, American Telephone and Telegraph Co., P.O. Box 8, Inverness, CA 94937; Station WLO, Mobile Marine Radio, Inc., 7700 Rinla Ave., Mobile, AL 36619; Station WOM, American Telephone and Telegraph Co., 1350 N.W. 40th Ave., Fort Lauderdale, FL 33313; and Station WOO, American Telephone and Telegraph Co., P.O. Box 558, Beach Ave., Manahawkin, NJ 08050.

Each station operates on upward of 18 channels. When calling the distant marine telephone operator, remember that much of their equipment is automated, and you'll need to make at least a 30-second call in order to raise their station on the first try. Giving your vessel name and call letters once won't do it. Call the marine telephone service by their call letters over and over again, repeating your vessel name and call letters many times and giving your approximate location. This gives the telephone station technicians time to select the best antenna for your incoming signal.

AMATEUR RADIO CONSIDERATIONS AND EQUIPMENT

Recent changes in FCC licensing procedures have made it easier to obtain an amateur radio ticket, and many cruising sailors carry amateur radio equipment aboard. Ham radio is not, however, a replacement for a marine SSB radio. A look at what ham radio can and cannot do for the mariner will clarify this frequent point of confusion:

- There are numerous maritime mobile "nets" through which ham radio operators who share similar sailing interests stay in touch. These informal nets get together at designated times of day and exchange weather reports, traffic lists, and other information, and shoreside ham stations are happy to help their counterparts far out at sea by "patching" a phone call to home through the land telephone lines or by ordering emergency parts. In fact, shoreside ham stations can be of great help in any emergency situation.
- Amateur radio operators engaged in local and

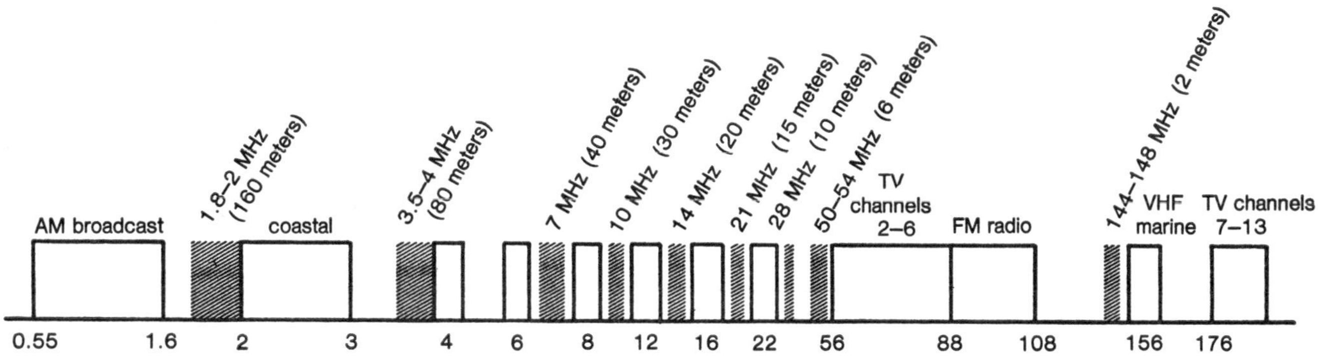

Ham radio frequency allocations.

coastwise cruising frequently carry 144-to-148 MHz VHF/FM radios (commonly known as "2-meter rigs") aboard. These sets are almost identical in size to the 25-watt marine VHF sets, and the ham VHF set can be tied into the marine VHF antenna system with a coaxial, 2-position switch. The ham operator can use his 2-meter set to call shoreside stations directly or through "repeaters" and get access to local phone patches. Local telephone calls patched in this way are free.

● If you're only looking for a way to avoid the marine operators on VHF or SSB frequencies and make cheap phone calls to your home or business, forget it. No business calls are allowed, and that includes the routine (as opposed to emergency) ordering of parts or supplies. Calls must be of a personal nature. Furthermore, ham radio is a self-regulating hobby, and the primary purpose of obtaining a ham license is to share the enjoyment of communicating with fellow operators worldwide. Shoreside ham radio operators aren't interested in being used strictly as telephone operators.

● The installation of amateur single sideband and VHF equipment is identical to that of marine SSB and VHF equipment. Since both types of systems use frequencies very close to each other, the elements of the installations are precisely the same. A ham single sideband set will use the marine SSB antenna system and frequently the same automatic antenna tuner. Only in an emergency, however, may a VHF or SSB ham set be used on marine frequencies. It is possible to modify a ham single sideband set to transmit on marine frequencies, and the Japanese have recently introduced a 2-meter ham radio handheld that will tune in marine VHF frequencies with a slight modification. Either practice is strictly illegal except in an emergency, and a ham could lose his license if caught operating non-type-accepted (under Part 83 of the FCC regulations) equipment on marine frequencies. In any case, you can now buy a type-accepted synthesized marine SSB set for the same price as a synthesized ham set. It is no longer true that ham equipment is much less expensive than marine SSB.

● As we have seen earlier in this chapter, synthesized SSB sets are now available with sufficient frequency capability to give you access to ham SSB frequencies (at 4, 7, 10, 14, and 21 MHz). When you have the appropriate ham license, and only then, you can transmit on these frequencies using your synthesized marine SSB set. Marine sets with ham radio capabilities are a good investment because the two services are almost identical in frequency, power output, and transmission mode.

The FCC has recently bowed out of the test-giving business, handing it over to the ham radio operators themselves. The FCC NPR Docket 83-27 outlines this volunteer examiner program in which local hams, many of them fellow mariners, prepare and administer the tests. And you can now study test guides that list, verbatim, every test question (and the right and wrong answers) that will appear on the examination you will take to become an amateur radio operator. The Morse Code requirements are still there, but it only takes about 60 days at 1 hour a day to master the International Morse Code at 5 words per minute, all you need for the Novice license. A fellow ham will give you the Novice exam, which consists of the code test and 20 easy multiple-choice questions.

The Novice license allows you to operate Morse Code on four worldwide frequency bands. You are allowed 200 watts of power output, and your communications range using code should be worldwide, day or night.

Most mariners want more than just code privileges, and that leads us to the Technician class license. It should take you no more than 60 days to prepare yourself for the 50-question Technician class theory exam. The exam is multiple-choice, and test guides are commercially available to help you study. No additional code test is required. The Technician class license allows you to transmit voice on the popular 2-meter band.

The General class license is the third step in amateur radio licensing, and this is the big one for mariners that cruise well offshore. The General ticket allows you to use single sideband voice equipment to talk on all worldwide amateur radio bands. The General class license requires only a code test—13 words per minute—and this will necessitate about 90 days of additional study. There is no additional written exam. If you don't pass the test the first time, you can take it again in 30 days from a local group of accredited volunteer examiners.

For information on amateur radio licensing and test-taking, contact the American Radio Relay League, Main St., Newington, CT 06111, or the W5YI Report, P.O. Box 10101, Dallas, TX 75207.

5

Radio Direction Finders

by Gordon West

Just below the AM broadcast band are frequencies allocated for radio navigation. The U.S. Coast Guard operates marine radio beacon stations that dot the coastline and islands in all U.S. cruising areas, and every major boating area and seaport worldwide maintains similar stations. In many areas of the world where loran signals are inaccurate or nonexistent (such as most of the Caribbean and the waters between Baja California and Panama), the radio direction-finding system provides a means of position finding and of homing into ports. These "point source" radio beacons transmit on frequencies between 190 and 410 kHz. All marine radio beacons are indicated on marine charts, and frequencies and broadcast schedules are given in the Defense Mapping Agency's publications No. 117A and No. 117B, *Radio Navigational Aids*. Some local marine electronics dealers give away frequency lists that include local radio beacons. Aeronautical radio beacons operated by the Federal Aviation Administration can also be used for direction finding; these are sometimes located on marine charts, or you can pencil in their precise locations if you know them. Aeronautical charts will show the locations of these beacons. Finally, commercial AM broadcast stations (550 kHz to 990 kHz) may be used for direction-finding purposes if the transmitters are advantageously located.

Two types of radio direction finders are available for tuning in "point source" stations on radio beacon and AM broadcast band frequencies. These are the portable or handheld radio direction finder (RDF), which can cost as little as $200 or as much as $400 or more, and the automatic direction finder (ADF), which will cost between $1,000 and $1,500. Both tune the same frequencies, but the ADF "fixes" on the point source station automatically, while the RDF requires you to do this manually.

With portable radio direction finders available for about $200, it makes sense to have one of the small units aboard even if you depend primarily on more sophisticated electronic navigational systems such as loran and SatNav. The latter have complex microprocessor circuitry and external antennas mounted in the weather, and on both counts they are vulnerable in ways that an inexpensive RDF set is not. And with its own batteries, a portable RDF will function even when the ship's power has been lost. There is no plan to phase out the radio direction-finding system; the Coast Guard will continue to maintain their radio beacons for the foreseeable future.

HOW IT WORKS

The U.S. Coast Guard operates two types of radio beacons—marker beacons that broadcast low-power signals from local harbors, and more powerful radio beacons that mark headlands (these are often located in lighthouses), major harbor entrances, drilling rigs, and other such features. The normal range for low-power beacons is 10 miles, while the high-power beacons can transmit their signals over distances of up to 150 miles, making them useful for position fixing by triangulation. High-power beacons are sometimes found in groups of up to six, transmitting sequentially in a timed pattern. Both types transmit very slow Morse code identifiers, making it easy for you to determine which beacon you have tuned in.

Radio waves between 200 kHz and 900 kHz (which includes the commercial AM broadcast stations as well as marine radio beacons) normally travel in straight lines radiating outward from the transmitter. In theory, all you need do is pick up the signal, determine what direction it's coming from, identify its source, and find the location of that source on your chart, and you'll have a line of position (LOP) along which you must be located.

The heart of a radio direction finder is its specialized antenna, which is highly directional to incoming radio waves. On a portable RDF the antenna windings are housed in a plastic bar that turns freely atop the receiver. Reception is strong when the bar is broadside to the incoming signal path, but when you swing the antenna end-on toward the signal source, the signal will disappear (a null). Because it can be located with much more precision than can the position of strongest signal reception, the null is used for direction finding. The unit is usually equipped with a meter to help locate the null. When it is located, the bearing is read on the 360° compass rose under the antenna. Either end of the antenna will elicit a null, so the user must make an independent determination as to which of the two possible directions (180° separated) the signal is coming from. This should rarely be difficult. If you are navigating on the West Coast, for example, a shore station signal is bound to be toward your east.

In the automatic direction finder (ADF) the principle is the same, but the antenna is a fixed double loop mounted outside the pilothouse, and signals are electronically "nulled out" while the twin loop assembly remains stationary. A pointer or digital readout will display the bearing of a signal. The ADF antenna also frequently incorporates an external sense antenna to resolve that 180° bearing ambiguity.

The procedures for using a portable RDF are simple. Let's say you want to home into a low-power radio beacon transmitting from a harbor. Swing your vessel into the approximate direction of that harbor, then tune in your RDF to the frequency listed on the chart. Doublecheck that the identifier you hear is indeed the marker beacon on the harbor entrance. Now swing the loop and look for the null. If the signal disappears precisely over the bow, you are heading toward the harbor, but if the signal disappears on the starboard bow, you need to alter course to starboard in order to home directly on the harbor entrance. If after a half hour of cruising on a steady course the radio beacon signal again appears a few degrees on the starboard bow, you are probably being set by wind or current. In this way an RDF can tell you how wind and current are affecting the accuracy of your dead reckoning. Obviously, these procedures will work just as well when you want to run a course directly *away from* a radio beacon, adjusting course to bring the "null" directly over the stern.

One of the most important uses for an RDF or ADF is for obtaining position fixes. Two or more bearings on chart-plotted point sources are needed for a fix, and radio beacons or commercial broadcast transmitters will enable you to get those bearings in low-visibility conditions, when you need them the most. Set the rotatable azimuth ring on your RDF to 000 in order to read the bearings in degrees relative to the ship's course, or rotate the ring to your current magnetic heading and read the absolute magnetic bearings. Try to find at least three stations, which should be in different compass quadrants; the wider the crossing angles of the bearings, the more accurate your fix. The stations might include a high-power beacon transmitting from a distant harbor on your port bow, a strong AM radio station with a shoreside transmitter on your starboard beam, and a small jetty marker directly astern of you. If the bearings are measured relative to the vessel's heading, convert them to absolute bearings in degrees magnetic. Plot the three lines of position (LOPs) from the charted transmitter locations. The LOPs should form a small triangle, the familiar navigator's "cocked hat." Your probable position is within this triangle.

There is also a simple radio direction-finding technique for checking speed over the ground. Find two shoreside radio beacons and on your chart draw a

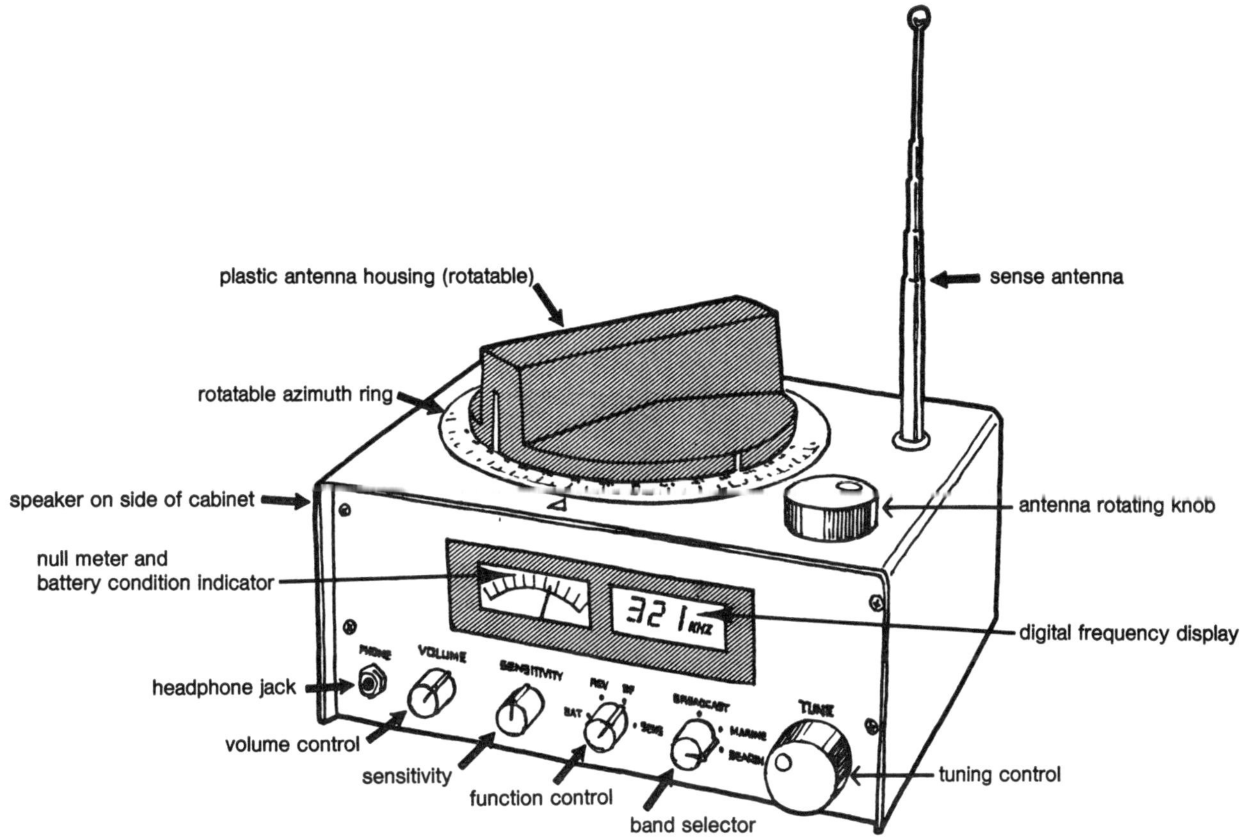

Typical portable RDF covering the radio beacon and AM broadcast bands. Some makes also have a beat frequency oscillator control, which helps in tuning a weak station by providing a squeal when the BFO is on to aid in finding the null. Others may include the 2-to-3 MHz marine band, but it is of little value for navigational purposes. The function control knob's four positions enable the user to check the voltage of the built-in batteries, to choose straight reception of AM broadcast stations (without any direction finding), to use beacon or AM stations for direction finding, or to energize the sense antenna. The price of a unit like this ranges from $200 to $500, and a typical size is 10 inches wide, 12 inches front to back, and 6 inches high.

line from each to intersect your course heading at 90°. Now measure the distance along your course between the two intersecting lines. Suppose that distance is 11 nautical miles. Start the clock when you are exactly abeam of the first beacon, and stop it when you are abeam of the second. If two hours has elapsed, your speed is about 5½ knots.

The accuracy of RDF bearings is seldom better than 2 degrees one side or the other, and frequently it is much worse. When stations transmit signals that must bend around landmasses or travel over hills or through valleys, the radio waves are likely to be redirected before they reach the ocean. For this reason it is important to know when the signals you are receiving from a radio beacon or AM broadcast station have traveled significant distances over an intervening landmass. These signals may have been redirected by as much as 20 degrees. This is one reason why marine charts don't list many aeronautical radio beacons or AM broadcast stations for direction-finding purposes.

Another characteristic of radio waves just below the AM broadcast band is that they may bounce off the ionosphere (an envelope of ionized gases above the earth) at nighttime, sometimes giving you highly erratic bearings to the point source beacon. These "skywaves" may approach from a direction as much as 35 degrees away from where the station is transmitting. Be extremely cautious of nighttime bearings on distant high-power radio beacons and AM broadcast transmitters. Within 40 miles of the transmitting station you are not likely to have any problem, but beyond that distance, groundwave signals may be overridden by skywaves, and this will give you a false bearing.

Despite the inaccuracies created by skywaves and coastal refraction of signals, a tremendous advantage of RDFs over other radio positioning devices is their simplicity of use. When you approach a fogbound harbor that has a radio beacon at its entrance, the accuracy of the bearings improves as the intervening distance shrinks. By the time you are within a few miles of the beacon, your positional error can be measured in yards. Furthermore, since you are homing on the beacon, it is only necessary to steer a course that nulls the beacon when the antenna is pointing dead ahead. No need to plot the displayed latitude and longitude on a chart, as you would need to do with a loran receiver, and from that fix, which could have considerable error in it, replot a course to the harbor.

Other, sometimes major inaccuracies can be introduced after the incoming signals reach your boat, but these you can control with the proper precautions. Metal surfaces and objects, for example, will deflect and distort radio waves. With this in mind, keep your RDF or ADF at least two feet above metal structures when direction finding, or better (since you can't very well operate your receiver two feet above the metal mast and rigging), develop a deviation card for the receiver in its normal position aboard the boat. The procedure is this: On a clear day, take a series of visual bearings on a local radio beacon (one on a jetty would be good) using a pelorus or a hand-bearing compass, and compare each such bearing to the corresponding RDF or ADF bearing. Three people—one on helm to swing the vessel, one to call off visual bearings, and one to take the RDF or ADF bearings—make the operation smooth. If there are major deviations, you may need to find a new location for the receiver. Metal stanchions can cause deviations of up to 10 degrees, and even the aluminum frame of a pilothouse window can redirect incoming radio waves. When you find a suitable and convenient spot, develop the deviation card for that spot. The operation is analogous to "swinging a compass."

When you use a portable RDF, which should be sitting on a tabletop or other convenient horizontal surface, make sure the set's lubber line is parallel to the vessel's fore-and-aft centerline. Naturally this applies equally to a permanently installed ADF. Just as important, make sure the helmsman holds a steady course while you take the bearings. Both these precautions are critical to the accuracy of the bearings.

You may need to shut down your engine to pick up weaker signals. The most common sources of interference include fluorescent lights, engine ignition systems, alternator whine, and electric motors.

WHAT'S AVAILABLE (AND BUYING TIPS)

There have been few advances in the technology of radio direction-finding equipment over the last few years. An RDF is really nothing more than a very sensitive radio receiver with a specialized antenna system. (When you're not taking bearings you can use the set as an entertainment radio.) About the biggest thing we have seen in the portable RDF is the switch from a slide rule–type frequency readout to a liquid crystal digital readout. This is a bigger improvement than it sounds, enabling you to tune in precisely the frequency of a very weak-sounding transmitter. There are times when the Morse code identifier of a radio beacon is garbled, but the signal is nevertheless strong enough for accurate detection of the null. With a precise digital frequency readout, you can be assured that you are indeed tuned into the right station. The old frequency scales could be a few kilohertz off, sometimes causing mariners accidentally to tune in the wrong station.

Most portable RDFs operate off flashlight batteries, and none are weatherproof. Keep RDFs and ADFs out of the moisture.

Some portable RDFs have an auxiliary "sense" antenna, the function of which is to resolve that 180-degree ambiguity. Given an accurate frequency tuner, an audible Morse code signal, and a local chart, however, you should have no problem deciding which of the two possible directions is the actual one. If the sense antenna comes on the unit you want, fine, but its presence or absence need not be a criterion of selection.

Handheld RDFs are available, these differing from the portable sets in that the antenna is housed in a pistol grip and the entire unit is rotated for null determination. A magnetic compass is substituted for the azimuth ring, which makes the unit easy to use but also introduces a new source of inaccuracy: The

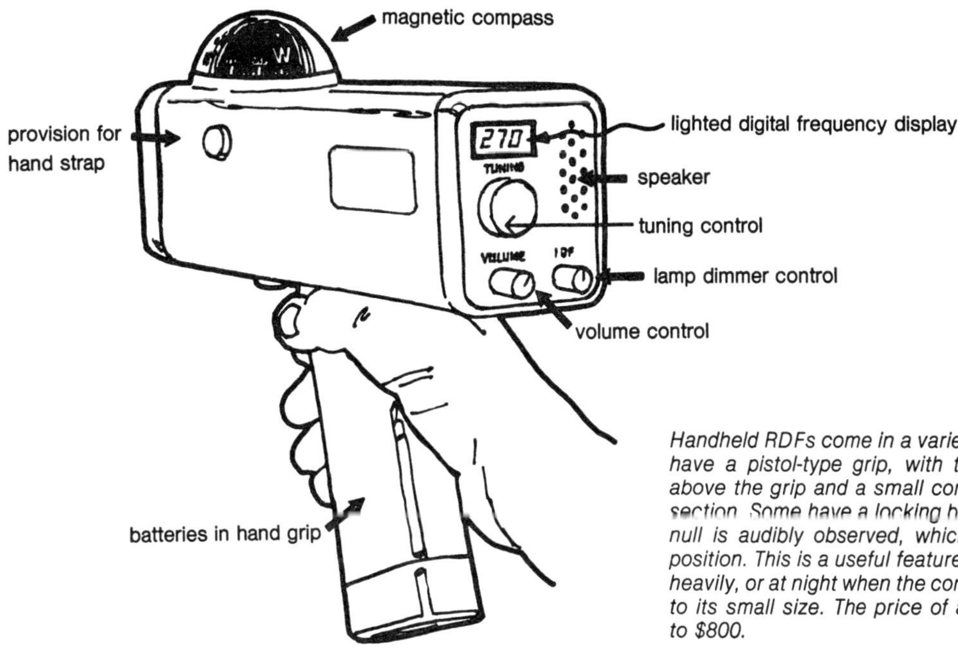

Handheld RDFs come in a variety of packages, but essentially they have a pistol-type grip, with the antenna and receiver controls above the grip and a small compass mounted above the antenna section. Some have a locking button that can be pressed when the null is audibly observed, which locks the compass card at that position. This is a useful feature when the boat is rolling or pitching heavily, or at night when the compass card may be hard to read due to its small size. The price of a handheld RDF ranges from $300 to $800.

compass is of course subject to magnetic deviation, and the small size of the compass card limits the precision of the reading. The handheld RDFs have their chief application aboard sailboats, since they can easily be taken on deck and held outboard, away from the standing rigging. They're priced in the same range as the portable units.

Most of the characteristics to look for in an RDF are similar to those that apply to VHF radios, as discussed in Chapter 3. An inexpensive RDF offering CB and FM bands (useless for direction finding) as well as the radio beacon and AM broadcast bands should be viewed with suspicion. Its "sensitivity" may be low, its "selectivity" poor, and its null may not be sharp enough for accurate bearings.

Sensitivity, the ability to pick up weak signals, is expressed as microvolts per meter, a smaller value meaning a greater sensitivity. The value should probably be 50uV/m or less on the radio beacon band, 40uV/m or less on the AM broadcast band. Selectivity is the ability to receive the desired station while filtering out interference from other stations, perhaps with stronger signals, on adjacent frequencies. Selectivity is rated in decibels—the higher the value, the better the selectivity. The value should probably be 30dB or higher. As with VHF radio, actually trying out the sets side by side in a marine electronics showroom is preferable to depending on quoted specs.

Automatic direction finders *can* be portable: The antenna is housed in a bar atop the set, as in the portable RDF, but a motor rotates the antenna automatically. Performance of these units is compromised by the motor, which can drown out the signal.

A permanently installed ADF with its large outside double-loop antenna can give you greater direction-finding range, many times greater than that afforded by a portable RDF. These units should be installed by a technician, since proper installation of the loop antenna's 8- to 10-strand wire is difficult, and because the units need to be calibrated when they are installed. Proper calibration increases the accuracy of the units, but again, a deviation card is necessary. These ADF units operate off the ship's batteries, so if you lose the ship's power, you lose the ADF. For this reason, and because a good ADF will cost considerably more than a portable RDF, an RDF is the more appropriate choice for an inexpensive alternative or a backup to loran.

A couple of additional things to consider when you shop for a radio direction finder:

Look for liquid crystal displays on RDFs and ADFs. The more traditional mechanical pointers tend not to be as precise or as troublefree. LCD readouts for both frequency and direction can give remarkable precision.

Look for an auxiliary connection on a portable RDF

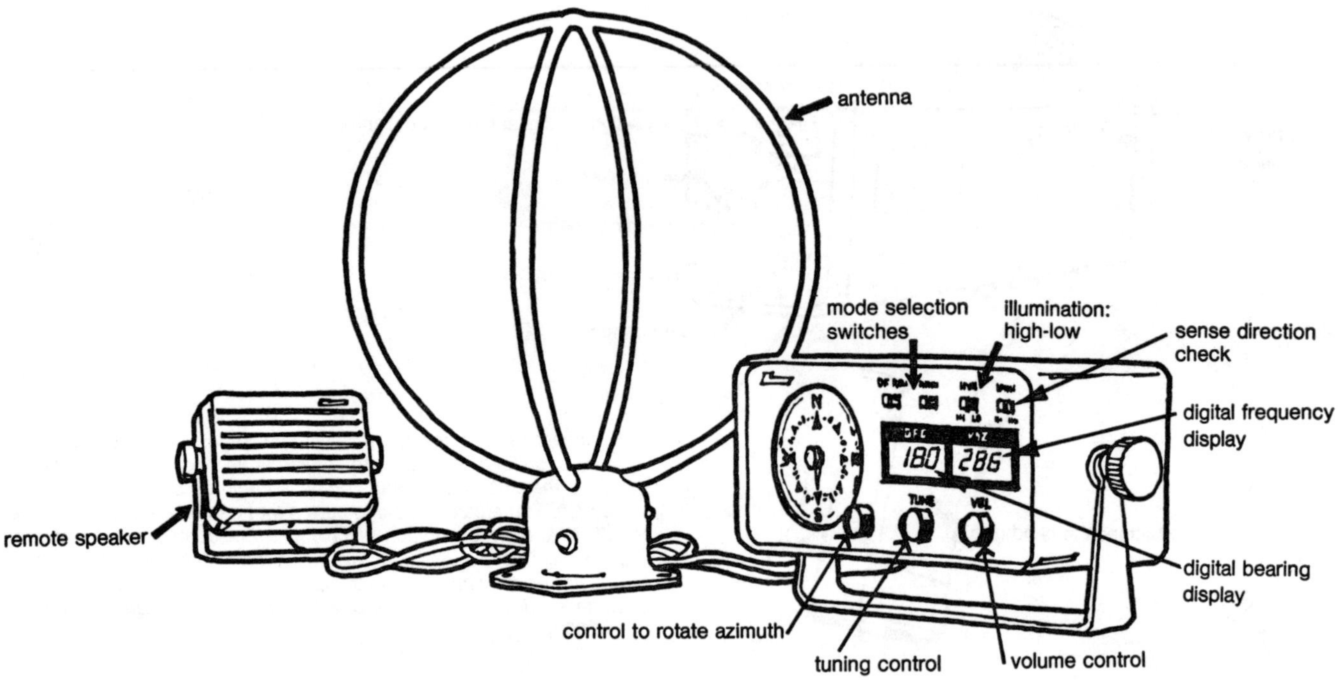

The two elements of an automatic radio direction finder are the topside double-loop antenna and the control and display console. The speaker may or may not be built into the console, depending upon make. The circular azimuth at the left side of the console can be rotated to the course being steered to provide magnetic bearings from the pointer over its face, or it can be aligned "north-up" to display relative bearings. There are also digital displays of bearing and frequency. Most ADFs include both marine radio beacon and AM broadcast bands; some also include the 2-to-3 MHz band, which is, however, of little use for navigational purposes. The price range of a unit like this is $2,000 to $4,000 with antenna, and a typical console size is 12 inches wide, 10 inches front to back, and 12 inches high. The antenna is typically 3 feet high.

to take an external 12-volt supply. This conserves the unit's internal batteries.

Look for a light on the RDF for night viewing.

An RDF is a prime candidate for catalog purchase. Almost all units are imported, and the catalogs usually offer far lower prices than marine electronics dealers. Most dealers won't work on the insides of defective RDFs, so why not buy through a catalog and ship the set back to the factory if it needs service?

If you buy an ADF and have it dealer-installed as part of the sales package, insist that it work properly and is properly calibrated, even if that means calling the technician back for one or more additional visits.

MAINTENANCE AND TROUBLESHOOTING

Always remove the batteries from a handheld or portable RDF before storing it. Old or dead batteries could begin to leak, damaging the battery compartment.

The biggest problem with portable RDFs is intermittent contact in the antenna assembly. The antenna windings housed within the plastic bar spin freely on tiny contacts that transfer the incoming signals from the antenna to the receiver section. These contacts usually develop small amounts of corrosion after a year or so of little use. The usual rotatable loop may be pulled off the RDF set and the contacts cleaned using television tuner cleaner (GC Electronics Tuner Tonic). After the contacts have been sprayed with tuner cleaner, the antenna is reassembled and spun many times.

VHF ADF

VHF automatic direction finders are the latest rage among fishermen because they can determine the direction of any incoming marine VHF transmission. If a commercial or sportfisherman locates fish and that information is disseminated over VHF channels, he may soon have company. VHF automatic direction finders are easily spotted by their unique four-pole

"Adcock" antenna system, which is comprised of four halfwave vertical elements electronically phased for making bearing determinations. Incoming signals are received by the four-bay antenna system and electronically mixed and matched, and the resulting bearing is displayed by a pointer needle, flashing light-emitting diodes, or a combination arrow and digital readout. If the antenna system is properly mounted high above all other antennas, bearing accuracy on VHF frequencies is usually better than 5 degrees. The units are obviously useful for search-and-rescue operations.

VHF signals from other ship stations cannot usually be received much more than 15 miles away. This is because VHF radio waves don't bend over the horizon, so that ships must be nearly within line of sight before VHF signals from one can be received by the other. This is why height of the VHF ADF antenna setup is very important.

Since some shoreside VHF stations operated by the Coast Guard or telephone companies have lofty perches, you can often tune in land stations over much greater distances, but few shore stations tell us where their transmitting antennas are actually located. A U.S. Coast Guard station in a particular harbor may use five or six transmitting antennas covering as much as 100 miles of coastline, and the local marine operator for your port might be transmitting from an antenna 75 miles outside the harbor limits. These remote antenna sights can complicate the use of VHF ADF for direction finding. These sets do *not* have the frequency capability to receive signals from radio beacons or AM broadcast stations. They are highly specialized for homing on other vessels.

Since VHF radio waves reflect quite nicely off buildings, billboards, metal masts, rigging, and even nearby VHF antennas, it may be impossible to home accurately on a distant ship station if you are in a harbor. You might "see" a ship station signal coming from shore rather than from sea. Once at sea, bearings will begin to settle down.

You can estimate the distance to a ship station by his signal strength on your VHF ADF set. If the signal is weak and plagued by static, the station is probably more than 10 miles away. If it's loud and clear, the station is within 5 miles.

The selection of VHF ADFs is large. For $1,000 you can buy one already built into a 78-channel synthesized 25-watt marine VHF set. The set operates like a normal VHF radio until you turn on the ADF portion, at which point it switches from the big

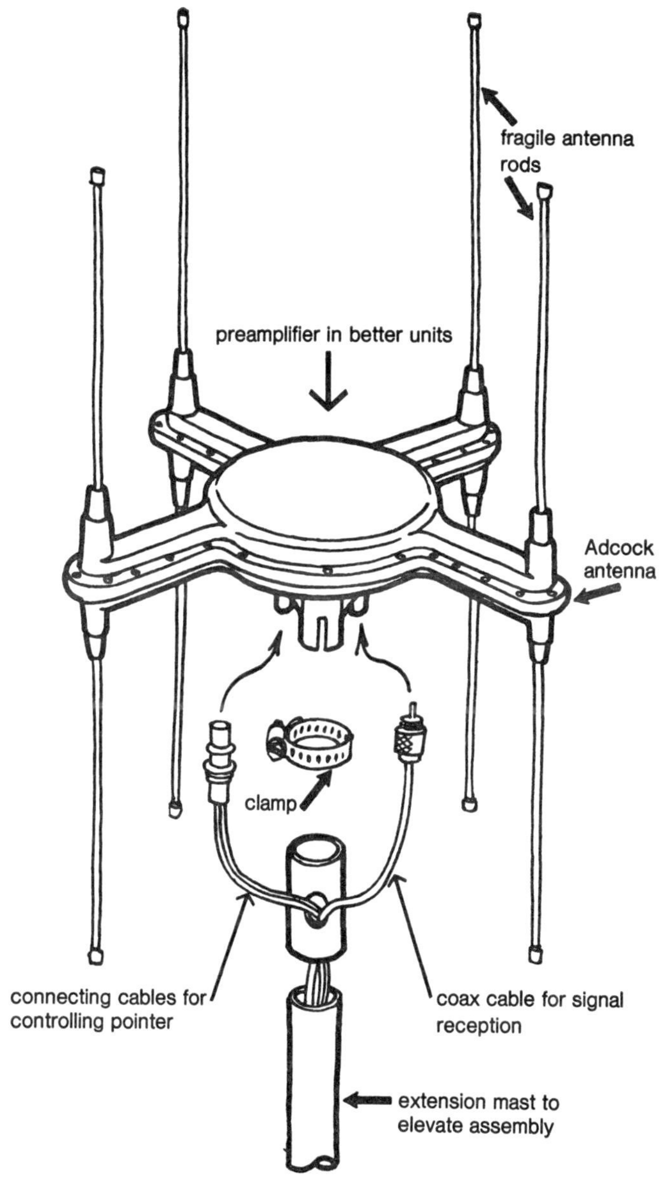

The antenna assembly used with VHF ADFs must be mounted above the boat's superstructure and away from other antennas. Alignment of the four antenna elements is critical, but easily done before securing the assembly to the plastic supporting mast.

transmit and receive antenna to the four-bay ADF antenna. The ADF antenna doesn't have the gain of a big white fiberglass whip, so reception will decrease slightly.

Another alternative is adding a VHF ADF assembly to your present VHF set. Regardless of what type or brand of VHF set you may already own, it's very easy to add VHF automatic direction-finding capabilities to it. These "add-on" devices normally cost around $500, and they are just as accurate as the built-in units. Audio from your present VHF set is taken out of the rear speaker jack and fed into your new ADF companion unit. The unit processes the audio through a phase comparison system while electronically "swinging" the four-bay antenna. It's a sensible way to go if you already have a good, synthesized, 25-watt marine VHF set installed.

These VHF ADF add-on units can also be tied into marine VHF handheld sets. The audio out of the handheld external speaker jack is fed into the ADF unit.

It's critical that the four-bay antenna be mounted well above all other antennas on your boat. Putting it down low will only cause incoming signals to distort, reflect, and refract. Some mariners might consider that weird antenna mounted high above all others on their boat a visual drawback.

When shopping for a VHF ADF inspect the antenna assembly. Are the rods hollow or solid? The hollow antenna elements will break easily, rendering your unit temporarily useless. The solid rods, such as those found on some of the newer units, will take more abuse and will break rather than bend. Be sure too that an extension mast is included with the antenna.

6

Loran

by Gordon West

Imagine yourself groping through the fog without benefit of radar, trying to get back to port with only a few gallons of fuel remaining. For the last four hours there has been no indication of land, no break in the weather—then suddenly, you pass a signpost firmly planted in the water that tells you where you are, which way to head for port, and the precise distance remaining along with an estimated time of arrival. Sound impossible? The signpost stuck in the bottom is a little farfetched, but a very sophisticated (and not so expensive) loran receiver will give you the same "waypoint" information. This type of capability is available in a small box that draws no more current that a running light.

"Loran" stands for long range aid to navigation. The system has been around since World War II, but not until the last 10 years has it enjoyed much popularity among pleasure-craft users. Loran manufacturers outside the United States have outdone themselves in producing smart loran receivers—and prices continue to fall.

In 1974, the U.S. government inaugurated Loran-C as the primary civil navigation system for the U.S. coastal confluence zone, covering navigable waters from the shoreline out to 50 miles offshore (or to the 100-fathom contour where this is beyond the 50-mile limit). The Great Lakes are also covered. Most loran signals travel well beyond the coastal confluence zone and can be used at distances up to 1,200 miles from the loran transmitting chain under some conditions.

Concurrently with the expansion and fine tuning of loran stations in the United Stations, Canada began improving Loran-C coverage on its east and west coasts. There is, as well, extensive loran coverage in the North Atlantic, Europe, the Mediterranean, Japan, Alaska, and Hawaii.

Loran makes possible precise and reliable position fixes. A typical fix in a strong signal area is accurate within a quarter mile, while on the fringes of loran coverage, the average accuracy is within 2 to 5 miles. The repeatability or precision of a fix is much better; you can return to within 25 feet of the exact spot where you lost an expensive anchor (or a person) overboard. It is this repeatability characteristic that enables bottom trawlers to avoid snags that have torn their nets in the past, lobstermen to find their buoys in the fog, and mariners to find in foul weather the navigation buoy whose loran coordinates they recorded when the weather was fair.

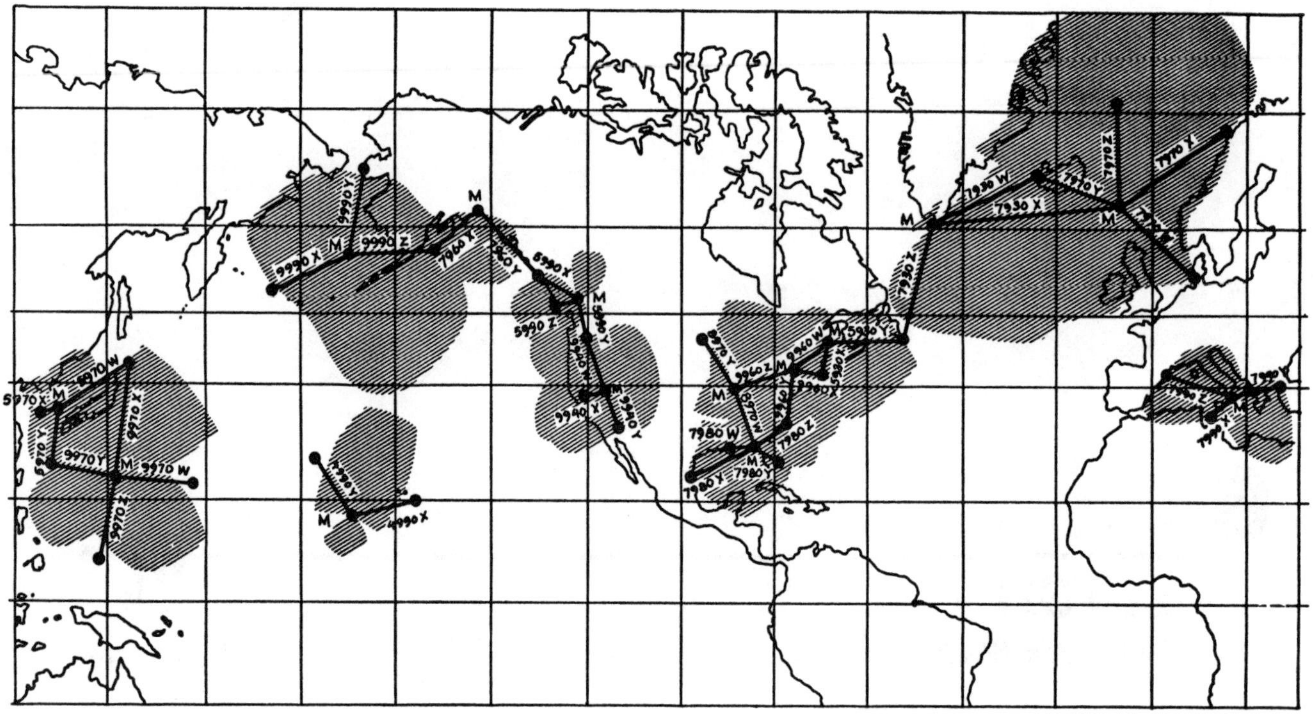

Groundwave coverage of the loran chains is denoted by the shaded areas. Ironically, the greatest coverage is in the extreme North Atlantic, where it is of little value to the average pleasureboat owner. The nominal seaward limit of the coverage as shown here is seldom realized.

HOW IT WORKS

We'll start with an oversimplified example: a man rowing his boat in the middle of a large harbor. At opposite ends of the harbor are two lighthouses that sound their horns every 30 seconds. Their short horn blasts are synchronous. When the rowboat is exactly halfway between the lighthouses, the man will hear both horn blasts simultaneously. If, however, his boat moves nearer to one lighthouse than the other, he will hear the closer horn first, because it takes the sound from the more distant horn a little longer to reach the man's ear. If the man knows the speed of sound and is nimble at math, he should be able to convert the time lag of the second horn to a corresponding distance. Then, having calculated how much closer he is to one lighthouse than the other, he can plot his line of position on a chart. A third lighthouse sounding its horn would enable the man to develop two more lines of position, and the point of their crossing with the first would give him a position fix. (Just one more LOP is all he really needs for this.)

Loran works on the same principle; the receiver measures the time difference (TD) between signals sent from a master station and secondary (or "slave") stations. In this case, however, the time differences are computed in millionths of a second (microseconds). Since radio waves travel near the speed of light, a microsecond of time difference is very meaningful to a loran readout. The TD between the master signal and the signal from secondary station X gives one line of position. The TD between the signals from master station and secondary station Y gives one more. (Loran receivers do not measure the TDs between two secondary station signals.) The crossing point of the two LOPs is the resultant position fix. Additional secondary stations providing additional lines of position can further increase the reliability of the fix, and most master stations are associated with three or four secondaries. The stations thus associated constitute a chain, each chain blanketing its designated area of coverage with short, timed pulses. These radio waves have a frequency near 100 kHz, giving them stable propagation characteristics.

A straight line drawn through a master station and one of its secondaries represents the "baseline" for the

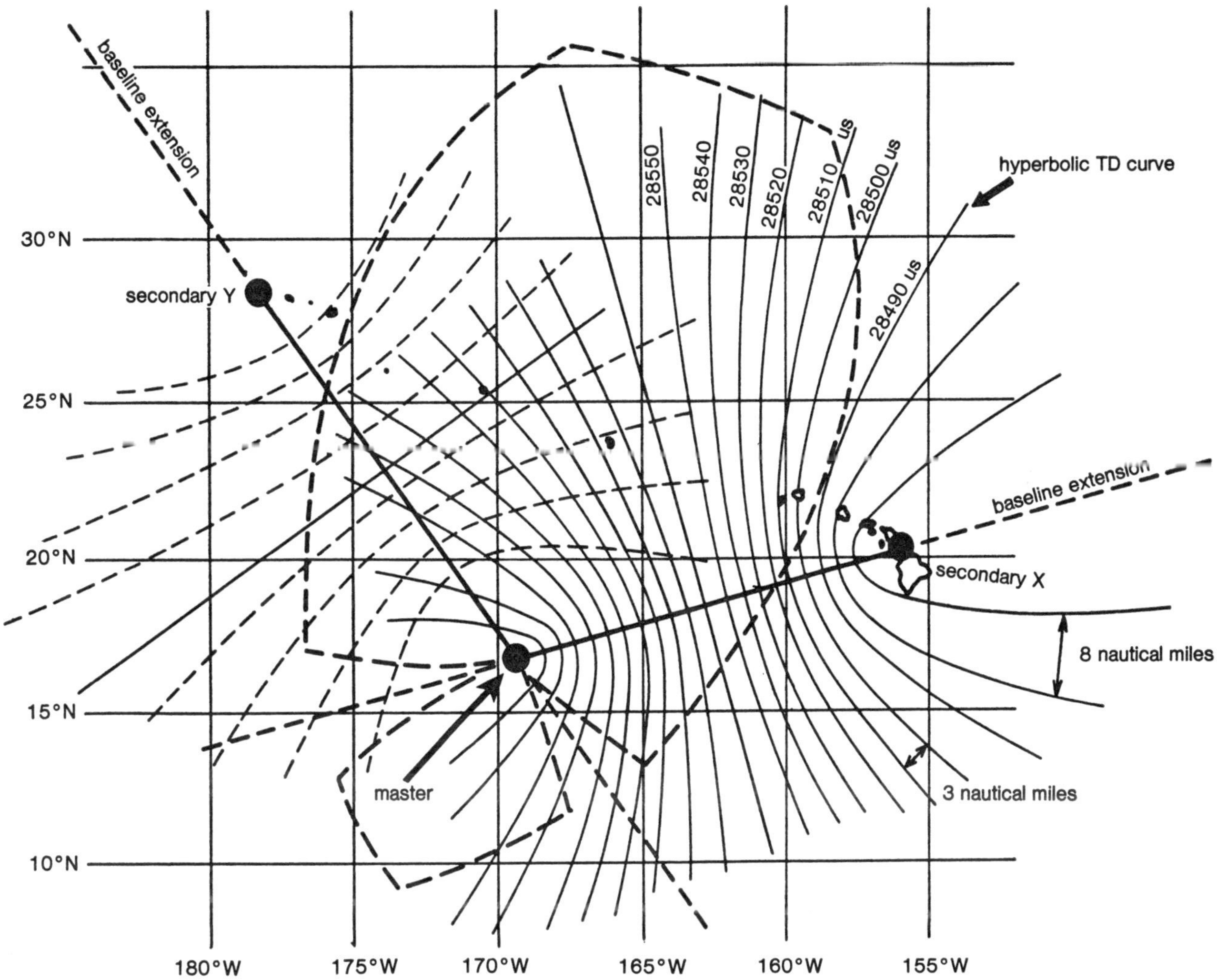

The Central Pacific loran chain, based in the Hawaiian Islands. The nominal coverage area is outlined by the heavy dashed lines. All loran chains have areas of better and poorer coverage with regard to crossing angles of the signals from the master and secondary stations. The gradient or distance between charted TDs also varies from one location to another within the coverage area, and this has a direct bearing on the accuracy of position fixes based on TDs. To make a plot using TDs requires practice and skill and is more time-consuming than plotting latitude-and-longitude positions.

two stations. If your course were perpendicular to a baseline and crossed it at an angle of 90 degrees, the time difference for transmissions from the master station and that secondary would be zero at all points along your course, provided our foghorn analogy is strictly true. Actually it isn't, quite: There would be signal interference if the master and all secondary stations transmitted at precisely the same times; therefore the pulse patterns transmitted from the stations in a chain are offset very slightly one from another. The principle is the same, however.

The master station transmits first, followed immediately by the series of secondary station transmissions from the chain. The time from one master transmission to the next is the "group repetition interval," or GRI for short. Every loran chain has its own GRI "radioprint" (like a fingerprint), and this lets a loran receiver identify the chain to which it is tuned. The desired GRI is either keyboard entered or selected or recalled from memory by the set and automatically tuned in.

Position fixing by means of loran is known as

"hyperbolic navigation" because the plotted lines representing given time differences between a master and one of its secondary stations are hyperbolic rather than straight (except for the line of points equidistant between the two stations). Unless you are within 20 nautical miles of one of the transmitting stations you are using, however, segments of these hyperbolas can be treated as straight lines for plotting purposes without introducing appreciable error.

Based on the fairly constant velocity of loran signals over water and the application of theoretical or measured correction factors where the signals must pass over land, loran time differences have been printed on nautical charts. These charted TDs are a good approximation in most cases, except where local anomalies were unmeasured at the time a chart was prepared. Although the accuracy of charted TDs can be expected to continue to improve as time goes by, you should be prepared for occasional discrepancies between what your loran receiver reads out and what your chart indicates for the TD coordinates of a particular position, discrepancies of one-quarter to as much as two nautical miles. Use the charted time differences only as an initial reference, revising select coordinates as you get to know and trust your own loran readouts for local waters. You'll want to doublecheck and verify them from time to time when the weather is fair (using the positions of charted buoys, for example), but the loran coordinates your receiver gives for particular locations in local waters should be highly repeatable.

Many factors can influence the accuracy of loran position fixes. For one thing, signals may propagate skyward, bounce off the ionosphere, and return as skywaves. Horizontally propagated groundwaves yield much better accuracy, and most loran receivers will precisely time the incoming groundwave pulses, automatically ruling out interference from a slightly delayed (though possibly stronger) skywave.

Whenever possible, secondary stations should be chosen to provide TD lines with large crossing angles, the nearer to 90 degrees the better. Fixes from such TDs will certainly be more accurate and reliable than those from TDs that intersect at less than 30 degrees. Some loran receivers will automatically select appropriate secondary stations.

The U.S. government has positioned loran stations to maximize crossing angles in the coastal confluence zone, but the angles may be too small for accurate position fixes in popular cruising areas such as parts of the Caribbean, the Southeast and Gulf coasts of the United States, and the waters off Baja California. Nevertheless, the repeatability of TD fixes made in these areas should still be quite good.

Another factor in the selection of secondary stations is the "gradient" of the TD lines from that station in your locale. The tighter the gradient (the charted spacing) between lines, the more accurate a position fix. A TD interval of 10 microseconds, for example, might be equivalent to one nautical mile or nine depending on location within the master-secondary coverage area.

Finally, if you are in line geographically with a master and one of its secondaries, but not between them (that is, if you are on one of the two "baseline extensions"), position finding using that secondary will be inadvisable or even impossible. A loran receiver that automatically selects appropriate secondaries would reject such a one.

Local weather conditions may also affect loran signals slightly. For this reason some loran chains feature computer-coordinated checking receivers that monitor the signals from the master and secondary stations. These reference monitors are located near important harbors where vessel traffic will be heaviest. Some loran chains will not only self-adjust their timing sequence, but will also notify your receiver when a secondary station may be experiencing a technical difficulty. This may register on the receiver as a blinking digit or a flashing light that flags the problem secondary. This "blink" alarm will be activated, warning you to doublecheck current readings with readings taken previously, if a loran transmitter gets out of tolerance by as little as 0.2 microsecond.

Probably the biggest culprit in occluding weak loran signals is interfering noise sources, either from the boat or from one of the many powerful military stations that transmit either side of the loran frequency band. Luckily, instead of getting a slightly incorrect reading in this case, the navigator gets no reading at all. Most receivers have built-in notch filters to help eliminate this adjacent-channel interference.

The majority of loran receivers in use today can, on demand, convert the TDs to corresponding latitude and longitude readouts, which many mariners find much easier and more familiar to use. The latitude/longitude readout is essential in areas, such as at the fringes of loran coverage, where the TD curves may not be printed on charts. If the conversion is made solely on the basis of the propagation speed of loran

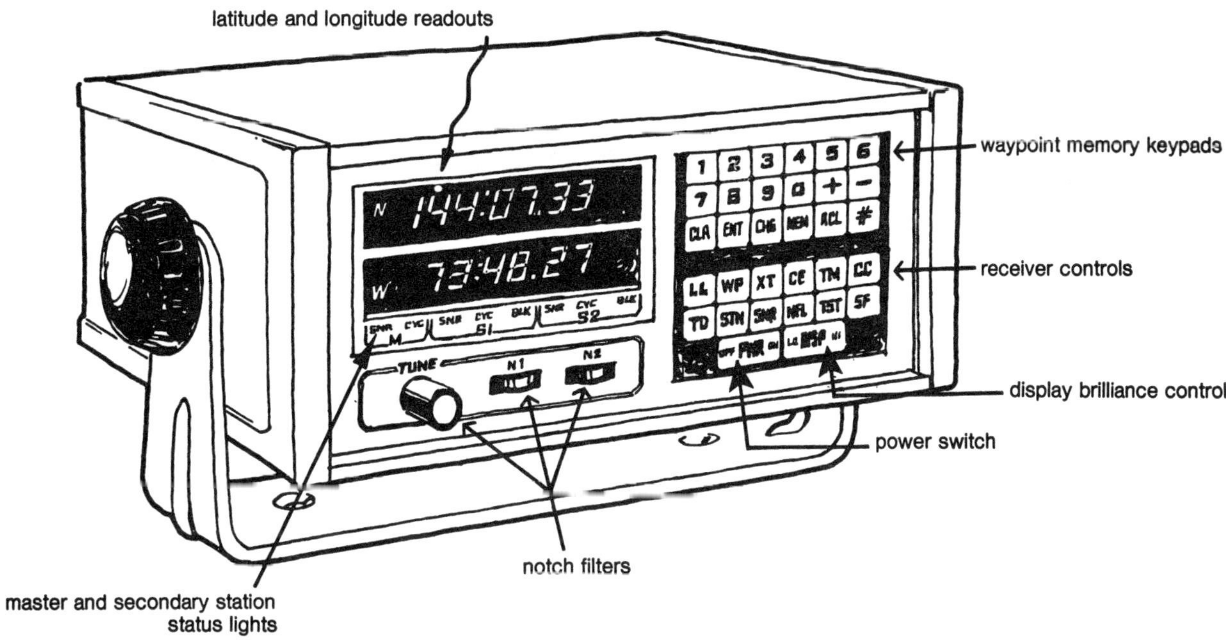

Two elements are common to all loran receivers: a display panel for either TDs or lat/lon readouts and a group of keyboard buttons to activate the various functions of the particular model. On the backs of the units are the connectors for ground, antenna, power, and (on some makes) interfaces with other electronic equipment such as autopilots, plotters, and recorders. Of great importance to the navigator is an operator's manual detailing the procedures that must be followed to utilize the many capabilities of the unit. The price range (with antenna) is $700 to $2,500, and an average size is 9 inches wide, 12 inches front to back, and 3 inches high, although there is little uniformity in size or styling.

signals over water, however, there will be losses of accuracy due to the landmass effect mentioned earlier in connection with the accuracy of charted TDs. Signals that travel over land are delayed slightly relative to those that travel exclusively across water; to compensate for this, a correction called the "additional secondary phase factor" (ASF) can be applied to the TDs. Some more expensive loran sets have factory-programmed ASF corrections; others require the user to keyboard enter the appropriate correction for the locale. There is also a seasonal influence when loran signals pass over land. In the spring, as trees leaf out, loran signals are delayed more than they are in winter.

WHAT'S AVAILABLE (AND BUYING TIPS)

More than 30 different types of loran sets are available today, the majority being imported, and all are good performers. Suggested list prices in the midrange are falling below $1,500. Mail-order houses are doing an excellent job of providing the most product for the least amount of money, but buying from a catalog gives you no chance to get the feel of the unit before you purchase it.

A loran set with the ability to convert TDs to a latitude/longitude readout is desirable in waters at the fringes of loran coverage or where the TD lines are not well charted. In Baja California and in the Bahamas, for example, the hyperbolic TD lines intersect at impossibly oblique angles, and it is very tough to interpolate and transfer the already-imprecise TD positions from one local chart to another. In areas like these, the latitude/longitude readout, while not, perhaps, highly accurate, is much easier to use and will be the only alternative when TDs are not charted. Unlike the accuracy, the repeatability of position fixes will usually be high in fringe areas.

As mentioned earlier, some loran sets apply ASF corrections automatically when converting TDs to a latitude/longitude readout, while others require the user to punch in the necessary corrections. The choice is a matter of personal preference, since the convenience of automatic correcting must be weighed against its expense (for the large amount of microprocessor memory necessary to store a sufficient number of local ASF factors) and the slight (no more than a few hundred yards) potential loss of accuracy it entails (because memory-stored ASF corrections must be averaged over fairly large geographic regions). ASF corrections to be entered manually can be taken from charts or from the *Loran-C Correction Tables*, published by the U.S. Defense Mapping Agency and available from chart houses. In areas not covered by these publications, mariners can develop their own ASF corrections by comparing latitude/longitude readouts with the charted positions of precisely located navigation buoys. A single ASF correction is likely to be accurate within a 20-mile-diameter circle, but there are local exceptions. Experience is the best guide.

All loran units are equipped with notch filters to screen out unwanted noise from frequencies near the loran 100 kHz band. A good set will have three or more, and in most receivers at least two of these are calibrated by the dealer to screen out known noise sources in the buyer's local waters. Be forewarned that a discounter selling by mail order, if it is not a factory-authorized dealer, may lack the facilities to "tune" a receiver for your locale.

Some receivers have, in addition to preset notch filters, other internal filters that automatically self-adjust to weed out unwanted signals on surrounding frequencies. Alternatively, there may be manually adjustable filters—either external or internal—that can be retuned for new cruising areas and new environmental or on-board noise sources. These may be desirable near large metropolitan areas or to combat high shipboard noise levels from engine ignition systems, fluorescent lights, bilge pumps, tachometers, etc.

The type of display is another variable among loran sets. Compared to gas discharge tubes and light-emitting diodes, liquid crystal displays are a little easier to see and getting more so. At night they are backlit for easy viewing, and they consume almost no power at all. The degree of resolution of the display also varies, many units displaying TDs to the nearest tenth of a microsecond, a few displaying them to the nearest hundredth of a microsecond. The latter might be considered overkill, implying more precision than even loran can habitually offer, since 0.01 microsecond is equivalent to less than 10 feet on most TD gradients.

Waterproof or weatherproof models are available for open-cockpit use. Some loran units are built in submersible cases and are ideal for open boats or small runabouts. Their pressure-sensitive keyboards keep moisture out, are made of a newly developed membrane that is not vulnerable to intense sunlight, and will probably last for years if kept clean and saltfree. These units may be portable, with a rechargeable battery pack.

It is also possible to purchase a unit with a waterproof remote-control "head," enabling operation of the receiver from the cockpit while the unit itself stays dry below.

Most if not all loran receivers are sensitive enough to receive the needed signals and obtain a position fix anywhere in the coastal confluence zone, and the repeatability of fixes is universally good. Blue-water voyagers and those intending to cruise extensively in the fringe areas of loran coverage should be aware, however, that there are differences among receivers in the ability to pick up a weak signal, separate it from unwanted noise, and use it as the basis for a fix. For those comparing manufacturers' specs, a "dynamic range" of 120 decibels (the norm is 110) and an "acquisition signal-to-noise ratio" of –18 to –20 (the usual is –10 to –15) might indicate a heightened ability in this regard (and a higher price tag), but probably there is no substitute for the advice of experienced users.

For most mariners, the matter of preference is decided by the front-panel controls—how easy is it to make the unit do its thing properly? Imagine that you are turning on a receiver and preparing to use it. Some units will require the operator to select and keyboard enter the GRI of the chain he wishes to use, while others will either use the last chain entered or automatically choose the best one for the locale. In the latter case the choice is made on the basis of signal strength levels, signal-to-noise ratios, crossing angles of the TDs, TD gradients, and the proximities of baseline extensions. These units will apply the same criteria in selecting the most advantageously placed secondaries within the chain. In essence, your thinking is done by the loran's microprocessor, a feature that is best appreciated when the navigator is in unfamiliar waters.

Once the unit is turned on, it may require 5 minutes to settle down and start giving consistent readouts. Some need less than 2 minutes, but it's usually best to allow at least five. In the interim, a more sophisticated unit will go through a self-testing mode while you passively wait. It will doublecheck every integrated circuit, perhaps telling you through its alphanumeric display in fairly plain English, "Testing receiver." If it has automatic notch filters it will examine the frequencies around the loran band and adjust the filters as necessary, perhaps telling you, "Testing and adjusting filters."

When the unit is ready, it will let you know with a beep or a flashing light. Usually the time differences appear first. Assuming the receiver has a latitude/longitude conversion, how easy is it to call up this alternative display? Some units do it with the push of a single button. With some you must manually punch in ASF corrections, as previously discussed.

Most loran receivers today do much more than position find. Just about all of them permit the programming of "waypoint" coordinates, a waypoint being a preselected location at which the navigator wishes to change course or make observations, or to which he wishes to return later. It is possible to reach or return to waypoints with a precision of 50 feet or better. Some sets can memorize more than 100 waypoints, and the number will continue to rise in the future. Let common sense prevail. While it may be handy to store a number of selected waypoints representing buoys or landfalls in your local waters, few mariners will use more than five or six waypoints in a single passage. The receiver may also give the bearing (possibly adjusted for local magnetic variation) and distance to the next or the last-visited waypoint when it is "asked" for this information by the pressing of one or two buttons. It may also tell you how fast you are going and how long it will take to get there, and it may beep when you approach or reach the waypoint. Some units will automatically inform you when and how far wind or currents are setting you off track for the next waypoint, and they will suggest a new course to steer; you can set a "cross-track" alarm to go off when you stray farther than 0.05 to 9.99 nautical miles off the desired course.

Other features are available—some useful, some more or less frivolous. These include clocks, alarm clocks, interval timers, and stopwatches. Some sets can be reprogrammed for waypoints at home, having backup batteries or memories that are not erased when the juice is turned off. Many units have a panic button that, when pushed, will store the precise current position in memory, enabling you to return to the spot where someone or something has been lost overboard.

With so many capabilities, sophisticated loran units frequently are not exploited to their fullest potential by their users. Some receivers are more "user friendly" than others, perhaps prompting the operator by means of messages on the screen to take the next step in entering waypoint information or requests through the keyboard. Manufacturers are sensitive to the fact that they have created something so sophisticated it may be scaring mariners off, and they are trying to humanize the equipment. One unit will flash "Baloney" if you try to enter a waypoint that's in the middle of a landmass, or "AYE AYE" in response to a keyboard-entered request.

Autopilot Interface

Many loran units will tie into an autopilot, and the National Marine Electronics Association has agreed on a standard 0180 autopilot output to make the interfacing easier. It is best to consult the services of a marine electronics specialist for this, but if both loran receiver and autopilot were manufactured after 1981, they probably have the appropriate inputs and outputs. Commercial fishermen can find this an invaluable tool.

To supplement the discussion above, here is a list of other features you may wish to consider when shopping for a loran receiver:

- Ability to reject skywaves in favor of groundwaves, if present
- Skywave warning indicator
- Loran chain false indicator
- Loran chain error indicator
- "Off the air" secondary station identifier
- Signal-to-noise ratio figures
- Low signal alert
- Permanent memory that is erasable only on special command
- Automatic sequencing of several displays
- Cross-chain interference rejection
- Convenient mounting brackets supplied
- Low power consumption
- Numerous service stations in case of repair throughout the world, which is a function of name brand
- Name-brand equipment with proven track record
- And, of course, competitive pricing

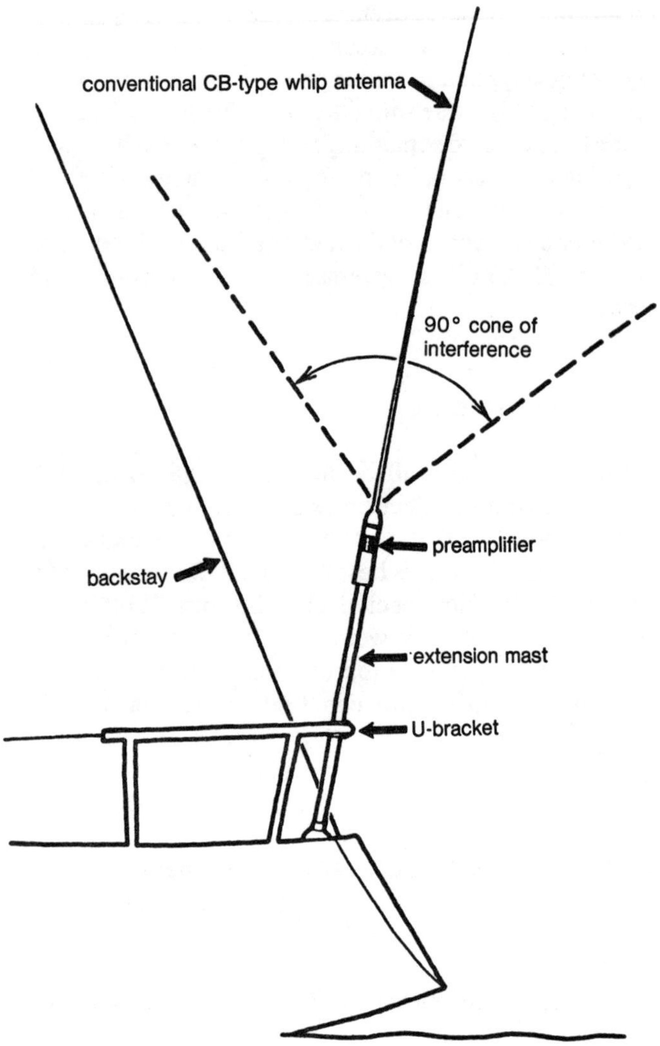

A transom-mounted loran antenna on a sailboat. The angle and height of the antenna is not critical as long as it is more than 45 degrees above horizontal. Regular CB-type antennas are satisfactory and relatively inexpensive. The coaxial cable from the preamplifier is fed through the extension mast to the receiver at the navigator's station.

INSTALLATION

A loran set is a very sensitive radio receiver without a loudspeaker; it can be installed by the boat owner provided proper care is taken. The necessary small current draw from a 12-volt source is easily tapped from the electrical panel. It is best to use an independent 12-volt line that goes directly to a circuit breaker, the breaker, in turn, having a direct connection to the battery. The 12-volt wiring doesn't necessarily need to be big—a pair of No. 12 marine duplex wires will work just fine. Wiring directly through a breaker (for safety) to the ship's battery system eliminates noise sources that might travel up a power lead from another connection point. A great deal of noise, for example, is generated in the chopper circuit of fluorescent light controls, and this noise can easily jump over to a loran power lead and cover up a weak signal. If tapped directly, the ship's battery will act as a giant filter to help "smooth out" random noise pulses found in other parts of the marine electrical system.

Since the loran receiver draws only about one-half to one amp, there is little worry that it's going to pull down your starting or auxiliary battery if you happen to leave your set on over the weekend. In fact, the current consumption is so low that you *could* power the set with a dry cell battery.

Like any radio receiver, a loran set needs a good antenna and a ground system, the latter to enhance weak signal reception and to help the receiver sort out weak groundwave signals from a high noise level. A good ground is also necessary to prevent dry-weather static discharges from erasing any of a set's waypoint memory. To get the most out of a receiver, a groundplane of 60 to 100 square feet is probably necessary. Components of this groundplane might include keelbolts, the engine block, through hulls, etc., as well as copper screen and copper foil when needed. Copper foil should be run from the receiver chassis and the antenna base to the groundplane. Methods of achieving a proper groundplane are discussed in more detail in the chapter on single sideband radio; while the procedures may be tedious and time-consuming, the standards of quality need not be as high for the receive-only loran unit as they are for a single sideband transceiver. An owner installation is therefore certainly feasible. (If you do intend to have a professional do the installation, buy the receiver from a marine electronics dealer and include the price of the installation in the purchase package. It will be less expensive that way than on a time-and-materials basis.)

It makes sense to use the antenna that a loran manufacturer recommends, particularly since most receivers are shipped with a matching antenna system. Most loran antennas, however, are nothing more than 9-foot white fiberglass CB whip aerials. The whip should be mounted out in the open and as far away from noise sources as possible. The heart of the antenna system is an encapsulated pre-amplifier, which the whip screws into. This pre-amp magnifies the incoming signal level and creates a narrow pass band for loran frequencies to enter. You may wonder how this low-noise pre-amplifier works without a DC voltage cable to provide power to it; current is indeed supplied to it, but that current uses the same coaxial cable that carries the incoming signal from pre-amp to receiver. One coax cable doing two jobs simplifies the antenna installation. If you do not know how to solder coax connectors, use the crimped, no-solder connectors discussed in the chapter on VHF radio. If you should ever break the whip, any CB radio–type whip will do quite nicely on the pre-amplifier mount. Most loran whips have a ⅜-by-24-inch thread, common among CB antennas.

Height is not important for a loran antenna; anywhere in the clear is fine. You can even rake the antenna back at an angle of up to 45 degrees without significant signal loss. Just keep it away from fluorescent lights, bait tank motors, refrigeration systems—anything that generates electrical noise.

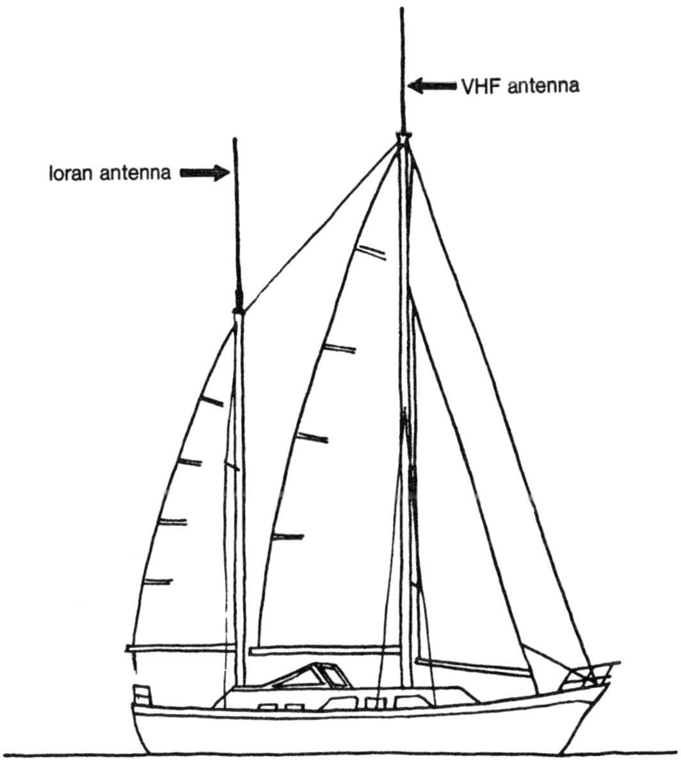

When power or sailboats have antennas for VHF or SSB equipment, the loran antenna should be mounted as far from them as is practical. Since any antenna is subject to damage, a spare whip should be aboard.

TROUBLESHOOTING

If your set does not function properly when the engine is running, noise-suppression steps are probably in order. Filters to eliminate ignition interference are available from marine electronics specialists, as are filters specific to just about any other noise source. Diagnose intermittent reception by having another person watch the display while you turn on and off possible noise sources.

Firm connections to antenna and ground circuit are essential. Any loose or corroded connections should be cleaned, tightened, resoldered, or replaced as necessary.

Nearby military transmitters can mask loran signal reception. If your receiver doesn't work in one locale but does work everywhere else, this is probably the problem. Factory or dealer adjustment of internal notch filters may be necessary. You may be able to avoid this by purchasing the unit from a dealer who is familiar with the interference sources in your local waters.

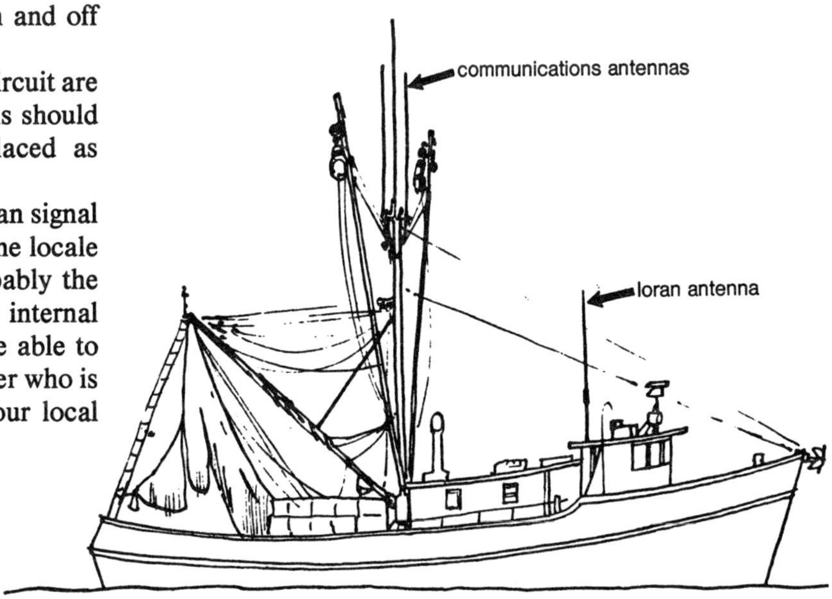

7 Satellite Navigation

by Gordon West

A satellite navigation receiver will display your position within 100 to 200 yards in any kind of weather—close to shore or thousands of miles out at sea. It will operate anywhere in the world. It displays your position as latitude and longitude, and there are no correction factors to apply (as there are for Loran-C). Prices have fallen below $3,000. The highly sophisticated microprocessor-based receiver is no larger than a VHF radio, and the antenna is also small. The satellites and associated ground stations upon which this navigational system depends will be maintained until at least the end of 1994, the United States Navy recently announced, and there will be five years' warning prior to the system's termination. (Its eventual replacement will be an even more impressive satellite array already dubbed the Global Positioning System.) If you buy a SatNav receiver in 1985 you can expect to get at least the better part of a decade's service from it, and probably more.

The satellite navigation system, then, has a lot to recommend it. No land-based system, including Omega and loran, can compete with the accuracy of a fix taken on an overhead satellite. Satellite navigation is the wave of the future.

HOW IT WORKS

The existing satellite navigation system, called the Transit system, was developed by the United States Navy in the 1950s to offer Polaris submarines a quick position fix with minimum surface time. These occasional surface fixes were necessary to recalibrate the submarines' on-board Inertial Navigation Systems. (Today, submarines use *both* Omega and the Transit system.)

After the Soviets launched the Sputnik I satellite on October 4, 1957, radio operators receiving its transmissions in the United States noticed a Doppler shift as the satellite approached, passed overhead, and then receded over the horizon. As the satellite was approaching, the transmitted frequency appeared several kilohertz higher than it actually was because of the velocity of the approaching "bird." This frequency offset disappeared momentarily as the satellite passed through the local zenith, only to reappear as a lower-than-normal apparent frequency when the satellite retreated. It was possible using this Doppler shift to formulate algorithms by means of which upcoming orbits and future satellite passes could be predicted.

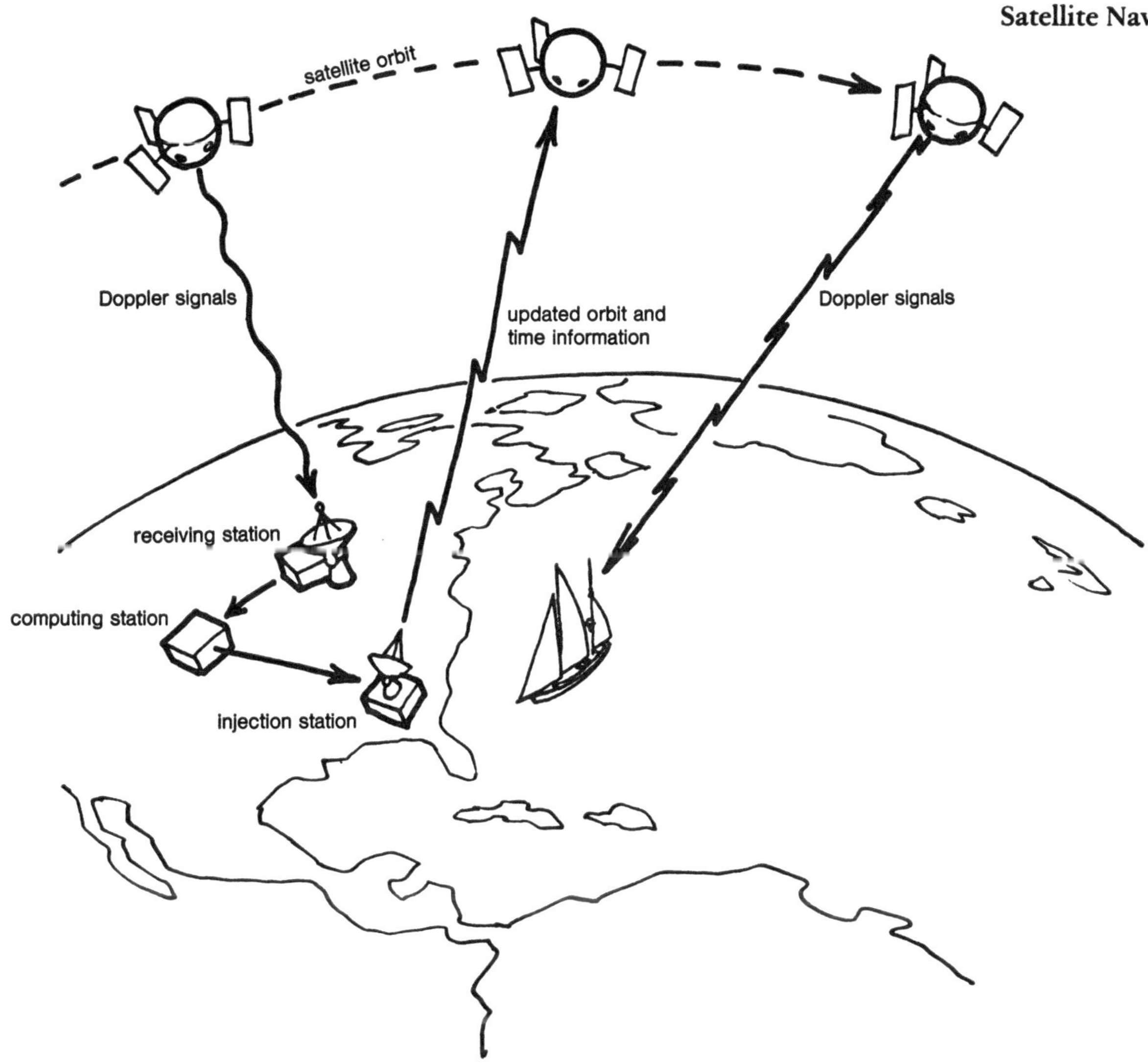

The satellites of the Transit system orbit around the north and south poles. Ground stations monitor the Doppler signals and periodically transmit to the satellites the precise and up-to-date parameters of their orbital paths. The satellites then transmit this information to vessels for position fixes anywhere in the world.

These algorithms proved extremely accurate, and they could be developed from Doppler shift measurements made at a single, fixed ground station.

By logical extension, if we know precisely where a satellite is, we should be able to use Doppler shift measurements as a means of fixing our position. This is the basis of the satellite navigation system. The satellite orbits are known and predicted, and with this information a SatNav receiver can tell us almost *precisely* where we are in terms of latitude and longitude.

The Navy first allowed the use of their Transit navigation system by civilian mariners in 1967. The system consists of five polar-orbiting satellites circling the earth every 107 minutes at an altitude of about 700 miles, their paths paralleling earth's meridians. This means the satellites are whizzing along at speeds in excess of 15,000 miles per hour. The satellite orbits are very close together where they pass over the poles, reaching their greatest distance of separation where they cross the equator. The orbits are fixed, creating the effect of a cage within which the earth rotates. Thus we see a different satellite on every pass. Although the time between local passes for position fixing is two to as many as six hours in equatorial waters, it varies between 35 and 95 minutes everywhere else. You can receive an updated position fix at least every hour in most of the world's cruising areas.

The five orbiting satellites are tracked by the Navy Astronautics Group in California, with associated tracking stations in Maine, Minnesota, and Hawaii. These stations monitor the orbits and periodically update the satellites' "on-board" navigational systems with commands transmitted from ground "injection" stations. Each satellite receives a new injected message about every 12 hours, and is thus always able to communicate its position with pinpoint accuracy. This system has been phenomenally successful over the past 10 years.

Every two minutes, each satellite transmits a navigational message intended for line-of-sight shipboard and land-based receivers. This encoded digital message contains precise orbital information that a SatNav receiver uses in calculating the satellite fix. While your receiver is picking up this two-minute "station break," it is also measuring local time and gauging the Doppler shift created by the approaching and departing satellite, correcting this for the slight Doppler shift caused by your own vessel's movement. Using these data, it computes a line of position from the satellite's known location. During the approximately 15 minutes the satellite is well above the local horizon, several LOPs are thus obtained and their best intersection determined to fix the vessel's position, which is displayed as a latitude and longitude readout on the face of your satellite receiver. The receiver gives you a beep to let you know when a satellite is beginning its transmission, and another beep 10 to 15 minutes later to inform you that all of the information has been processed and your position is displayed on the LCD readout. Although the satellites have been in continuous use for the last 10 years, the five "birds" are performing nearly flawlessly. There are backup satellites waiting to replace any that might fail.

The satellites broadcast on radio frequencies in the VHF and UHF region using a tiny one-watt transmitter. Their huge antenna systems, which point down toward the earth, intensify outgoing signals. Internal batteries charged by solar cells keep the transmitter going 24 hours a day, and an elaborate system of counterbalancers keeps the satellites' directional antenna pointing always toward earth.

Local weather conditions, day or night ionospheric variations, sun spots, solar disturbances, and radio "blackouts"—none of these have any effect on the incoming UHF radio waves. About the only thing that can adversely affect SatNav reception is an impediment or obstruction in the line-of-sight path between your SatNav antenna and the satellite. On some passes the bird may be barely above the local horizon, and reception will be weak. In this case, you will not get pass information, and you must wait until the next satellite comes around in another hour or so.

It has been found that the satellite transmission can reach your small (about 3 feet long) satellite antenna even if someone is standing in its path. A metal object between the antenna and the orbiting satellite, however, might cause it not to "see" the incoming signal, so don't dry out a dish pan or metal fishing lures by putting them over the antenna. Try to put the antenna out of the way, but don't worry too much about a sailboat's aluminum mast—most satellite signals will filter through the rigging and around the mast and can be collected by the antenna quite nicely. Mounting the antenna too close to a powerboat's steel stack might reduce operating capabilities. The height of the antenna is not critical. Once the antenna is installed, simply keyboard-enter the antenna height into your receiver's microprocessor memory.

Between satellite fixes, the SatNav receiver displays a continuously updated dead reckoning (DR) position. This requires that course and speed information be entered into the receiver. (The receiver also needs this information in order to make accurate computations from Doppler shift measurements during satellite passes.) If you are on a steady course and speed, you can enter the information manually via the receiver's keyboard, and the receiver will give you, in addition to an updated DR position readout, the precise time that you can expect your next satellite position fix.

A far more efficient and accurate method of inputting speed and heading information for a receiver's DR calculations—especially on sailboats, for which steady speeds and courses are the exception rather than the rule—is to use a fluxgate compass and a speed sensor that are electronically coupled to the receiver. You would do well to use the heading and speed log systems recommended by each SatNav manufacturer. While it is possible to interface a new SatNav receiver with your present speed measuring system, the attempt may cause a lot of headaches, and it will complicate the installation process. Find out ahead of time whether the SatNav receiver you are thinking of buying will interface readily with the speed log you already have. Most satellite manufacturers produce their own fluxgate compasses that mount almost anywhere for heading information. Use the

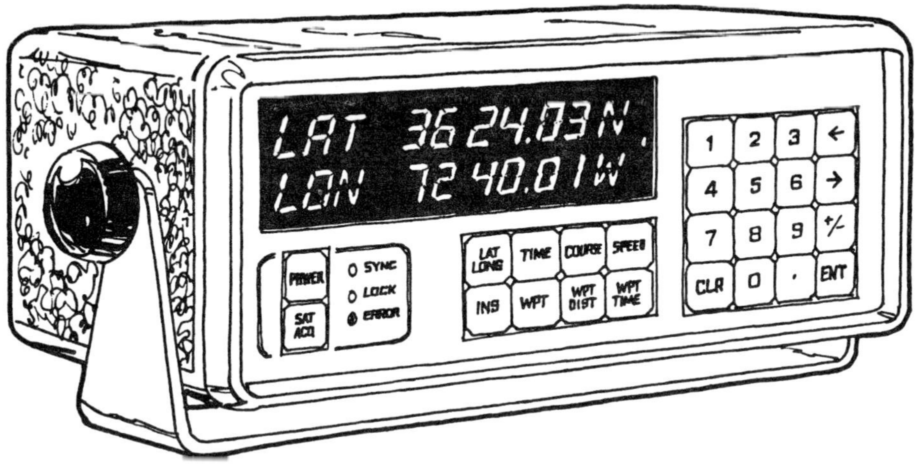

Satellite navigation receivers display the current latitude and longitude, which can be plotted on the appropriate chart. Using the keyboard, the operator can prompt the receiver to display many other types of navigational information, such as distance to a destination, course to be steered, speed made good since the last fix, total distance traveled, and even the time within one second. The price range (with antenna) is $1,000 to $5,000, with the more expensive units being capable of tying into such instruments as an autopilot, knot log, and compass. The size of a typical unit is 12 inches wide, 8½ inches front to back, and 4 inches high.

compass offered by your receiver manufacturer. Once you have these electronically coupled sensors in place, your SatNav receiver will receive continual updates on heading and speed changes.

The SatNav receiver's DR calculations are subject to the same inaccuracies—especially those caused by currents and leeway—that affect all DR plots, the difference being that with each new satellite fix the slate is wiped clean. In addition, inaccurate heading and speed inputs can cause errors in the receiver's position-fixing computations during satellite passes. An error of 1 knot in the estimated speed over the bottom, for example, will produce an error of as much as one-fifth of a nautical mile in the computed position. On an ocean passage, of course, such an error matters little.

WHAT'S AVAILABLE (AND BUYING TIPS)

Plan on spending $1,000 to $5,000, for which you will get some incredibly small equipment that will do an outstanding job. (The fluxgate compass, if one is available for your receiver, is not included in the basic price.) All the receivers on the market are of good quality, since uniformly high sensitivity is necessary to receive the satellites' one-watt transmissions. All the receivers execute their fundamental task—position-fixing during satellite passes—with efficiency. The less expensive receivers merely have fewer optional features. New, top-of-the-line satellite receivers feature waypoints, anchor watch alarms, and all of the amenities found in sophisticated loran sets. Some satellite receivers will even tie into a loran set to give you additional information between satellite passes, although you will probably be disappointed in the loran output once you get spoiled by the accuracy of satellite navigation. Indeed, some mariners see no need for loran at all. They consider a depthsounder, an RDF, and a radar unit to be a sufficient electronic navigation package for coastal waters, with a SatNav receiver being the next most desirable item.

Choose a system that's easy to understand and operate—and most are. The controls have been made "user friendly" with display prompting—most readouts will ask you what you want next. Unless you have automatic speed and heading input, the readout will also ask you for that information. The amount of information a more sophisticated receiver will display is dependent on its programming. If you elect to program your unit's microprocessor memories to the hilt, it will give you a veritable wealth of output. If you let the receiver function on its own, it will simply tell you your position within a few hundred feet every hour or so, marking the event with a friendly beep.

In most cases, no tuning is necessary on a new

receiver, and installation is easy unless you encounter a speed log–receiver interfacing problem. At the manufacturers' request, SatNav receivers are usually repaired by the manufacturers rather than the dealers, at least when the malfunction involves the integrated circuitry. Even if you buy your unit from a specialty dealer, there is a good chance it will be returned to the manufacturer for repair (although the dealer *may* give you a loaner unit to use until yours comes back). For all these reasons, you might consider buying your SatNav receiver from a discounter and saving some money, *if* you have a good idea of what you want.

INSTALLATION

Installation of the equipment is more than straightforward—in most cases it's a snap. Noise levels and grounding are not an important matter in most satellite installations, so grab 12 volts wherever you may find it on your main circuit breaker panel. Just make sure that it's a good solid 12-volt source, free from alternator spikes or battery charger hums.

If you have good grounding throughout your boat, run a small ground wire from the chassis of your satellite receiver to any convenient ground point. This cuts down on the amount of radiated microprocessor noise that most receivers generate—noise that can sometimes cause problems in the use of a weather facsimile receiver, ham rig, or marine SSB transceiver. Satellite receiver manufacturers have been slow to admit that their microprocessor circuits are creating electrical noise aboard boats—but most are finding better ways to shield their receivers' innards.

The antenna system is fed with coaxial cable. As with VHF radio and loran systems, more and more manufacturers are enclosing crimped, no-solder coax connectors for do-it-yourself installations. These work well, though they do not last as long as soldered connections. Crimped connections are available separately if they do not come with your unit. Most antenna systems feature a pre-amplifier, similar in style to the loran pre-amp. This pre-amplifier is at the base of the antenna system and picks up its voltage from the same coaxial cable feedline that carries incoming signals from the active antenna element. The antenna is usually sealed in white fiberglass. Mount it as near vertical as possible for maximum reception during low satellite passes. You can, if you wish, hide the antenna belowdecks in a fiberglass hull and still get adequate satellite reception. Mounting the antenna high atop a mast is totally unnecessary, and is really not recommended.

Interfacing the receiver with a knotmeter will be the only possibly difficult aspect of the installation. Here a dealer can be a great help, saving you from possible frustration. If you want the knotmeter interface, and you want to go it alone, you can contact the manufacturer or seller of the receiver and obtain the appropriate interfacing kit. You'd be better off not to go this route, however, if you're not good with a soldering iron.

THE GLOBAL POSITIONING SYSTEM

As things stand now, the Global Positioning System (GPS) will be instituted sometime in the 1990s, and the satellites are already being placed in high-altitude orbits for testing. The GPS plan calls for 18 satellites in orbits as far out as 10,000 miles, as compared to the 700-mile altitudes of the Transit satellites.

The GPS satellites will always be in view of vessels everywhere, so that fixes may be obtained every second. There will be so many satellites within view at any one time that they will create a three-dimensional effect, and elevations as well as positions can be calculated. The system will be a natural for airplane pilots, land survey teams, and other users that want to know altitude as well as latitude and longitude.

The GPS will not be made available to civilian users for some years. When it is—if you want to pay the price (maybe $300 per month)—you will be able to buy equipment that will exchange communications with the satellites for position fixes as precise as 6 inches on any landmass. You will probably be able to use the system for free and enjoy position fixes within 30 yards. Close enough?

But for most of the next decade at least, the present Transit system will continue in full operation, so don't be concerned that your SatNav receiver will be useless or obsolete within a year.

8 Autopilots

by Freeman Pittman

Autopilots are Elmer Sperry's gift to pleasureboaters, or they are a pain in the neck that never work when you need them. People's reactions to them are based on experience, and when the system is a good one, you don't forget it. The sensation of feeling the boat moving underfoot surely and economically, with no one at the helm, makes you glad you were born into this technological age to witness such feats of intellect over nature. When the experience is a bad one, a good deal of the unhappiness stems from not knowing enough about what you're getting into before you purchase and install the unit. And what you need to know is significant, because the autopilot is possibly the most complex system on the boat, and takes up a good deal of space and energy. Discovering the strengths and weaknesses as well as the idiosyncrasies of each model that interests you will go a long way toward ensuring the reliability of the pilot you choose—and your ultimate happiness.

Autopilots have grown immeasurably in sophistication in recent years, and yet the most basic design of more than 15 years ago—the Tillermaster—is still being marketed successfully, offering you dependable, efficient operation along the coast or while you're crossing an ocean singlehanded. The least expensive electric pilot on the market costs $500, much less than loran and about the same price as a handheld VHF, a cost many small-boat owners can justify.

On the other hand, the most thorough, permanently installed belowdecks system can keep you on a compass- or windvane-directed course with an accuracy of plus or minus 1 degree in good weather, compensating for yaw, leeway, or an imbalanced helm in sailboats, and even for current set and drift when a loran is tied into the system. It will do so while using little more of a sailboat's precious electrical power than your cabin light does. Powerboat systems, by the same token, deliver more power from lighter, more compact systems than ever before. Trim, torque, and poorly synchronized engines can be offset by these pilots, too.

HOW IT WORKS

Years ago the crudest autopilots simply zigzagged the boat across a chosen course, using electrical contacts touching the heading sensor's compass card as the limiting switches. Because of their behavior, these basic pilots have been dubbed "hunting" types.

The improvements that made autopilots viable for small craft were miniaturization of components and development of highly efficient electronics during and after the Second World War, along with the incorporation of a "deadband" in the system. The deadband, whether mechanical or electronic, is a narrow segment of the compass down the middle of which the boat's course is set. As long as the boat's heading stays within the deadband, say, 2 degrees on either side of the course, the autopilot takes no corrective action. The total play in a manual steering system is a deadband, too. Wiggle the steering wheel a half inch or so, and you feel no rudder movement. The boat is free to wander within this range. Obviously, allowing the boat some free rein saves energy. Within the typical 1- to 5-degree deadband width current autopilots allow, the boat, if it has been trimmed and balanced, will come back to the desired course on its own accord a fair amount of the time. Making the autopilot work only outside the deadband lessens the amount of electrical power needed to drive the motor that turns the rudder.

A higher level of sophistication, if not necessarily a more cost-effective one, is reached by "proportional-response" autopilots. Proportional response means the rudder is turned at a particular angle and rate for a given course error. Less rudder is needed for a small error, more for a large error. The proportional pilot applies the right amount for each, unlike a simpler system, which reacts with the same amount and speed of helm correction for all course deviations and conditions.

Regardless of the principle of operation, an autopilot works only on the command of its lookout. For the powerboat and (most of the time) for a vessel under sail, the lookout is the "heading sensor." For many years, autopilot systems have incorporated a standard magnetic compass as the heading sensor. An electrical coil may be placed above or below the compass card and rotated to align itself with the field created by the compass's magnets. Then, as the coil (and boat) rotate out of alignment with the surrounding field, a low-voltage signal is inducted in the coil and sent to the control unit for amplifying and passing on to the drive unit.

Another effective sensing system uses a photocell to detect a light beam through a compass card divided into clear, partially shaded, and opaque sections. To choose a course, you turn the card manually on the simple cockpit-type pilot; a motor will turn it for you in the permanently installed systems. Then line up the boat so the light beam passes through the translucent part of the card. If, through rotation of the card caused by a course deviation, either the opaque or clear portion swings into the path of the light beam, a signal is generated that activates the pilot's drive motor.

These systems are all repeaters; they follow the movements of the boat around the compass card and detect the differences. Many manufacturers today are turning to a fully digital heading sensor that replaces the card compass itself. This is the "fluxgate compass." You don't see numbers on a card with the fluxgate; instead, its inner works resemble the antenna of a tiny radio directionfinder. It works much like an RDF, too; the radio signal is strongest when the radio directionfinder's bar antenna is at right angles to the signal and weakest when it is end-on to the signal. For the fluxgate, consider the earth's magnetic field as the radio beam, and use two bars crossed at a right angle as the antenna. Pass a low voltage through coils wound around the bars, and watch the digital readouts as the earth's magnetic field distorts the coils' voltage output. The amount of disturbance varies as the fluxgate is rotated through 180 degrees. The phase changes too, and sensing the phase eliminates reciprocal ambiguities. A given voltage output and phase correspond to a signal assigned to a particular heading on the compass rose, which is transmitted to the autopilot control unit for amplification. The system is much less complicated mechanically than one based on a card compass and repeater. Power is saved as well. The only problem with the fluxgate compass is that it appears to be more vulnerable to temperature variations and to dip error than a card compass, which can be fairly easily weighted to deal with dip occurring in the temperate and arctic latitudes. Autopilot manufacturers are jumping on the fluxgate bandwagon nevertheless.

Some manufacturers—a surprisingly low number—offer a windvane option as an alternative to the heading sensor for sailboats. It is just like an electronic apparent-wind direction indicator carried at the masthead, and you dial in the relative wind angle you want to sail, usually close-hauled. The wind-direction sensor takes over from the compass sensor, and the boat will stay on a heading relative to the wind, not a compass course. This is most useful while sailing either on the wind or dead downwind.

A heading error signal from compass or windvane goes to the control unit for processing and amplifying into a signal strong enough to command the actuator motor. The control unit may be better envisioned as a control group; it performs a number of functions, and not all of its components necessarily reside in one housing.

As to understanding what all the functions are, you cannot rely on the manufacturer's literature to help you. Manufacturers refuse to standardize their terminology. The following explanations should help to resolve confusion.

Yaw (called variously seastate, sensitivity, response, or deadband) sets the deadband width on pilots using that type of control; that is, it determines the amount of deviation from course (the yaw angle) the pilot will allow before making a rudder correction. The control may be a mechanical one inside the pilot that you set once for your boat's characteristics, or it may be a dual adjustment you can make at any time to suit the sea conditions. A wide deadband lets the boat react to momentary influences, such as individual waves, in heavier conditions. A narrow deadband keeps the boat on a straight course in smooth water. Given heading sensors with plus-or-minus 1-degree accuracy in smooth conditions and drive motors able to make small corrections with minimal time lag, today's autopilots offer a deadband as narrow as 2 degrees (1 degree to either side of the course) and adjustable to as wide as 5 to 8 degrees. Autopilots maximizing the automation their microprocessor logic allows do away with the deadband control by feeding a nearly constant stream of small power pulses to the drive unit instead of fewer but larger bursts in response to wider yaw angles. For these pilots, a low-pass filter, which won't allow momentary errors from individual wave impacts to be signaled to the drive unit, offers the damping inherent in deadband systems.

The *gain control* (rudder, ratio, P-factor) adjusts the degree of rudder response to course error. For a given error on a light sailboat with detached spade rudder or a planing powerboat, the gain might be set low, since little rudder movement is needed to make the boat change course. On a heavy, long-keeled sailboat with attached rudder, or a trawler-type powerboat, more helm might be needed to make the boat respond. In this case, a higher gain setting makes the rudder turn farther for the same degree of heading error. In a proportional-response pilot the gain may also control the speed, or rate, at which the rudder is turned in relation to the rate at which the boat falls off course. From this relationship comes the term proportional rate.

Essential to a proportional-response system is an indicator that signals rudder position to the control

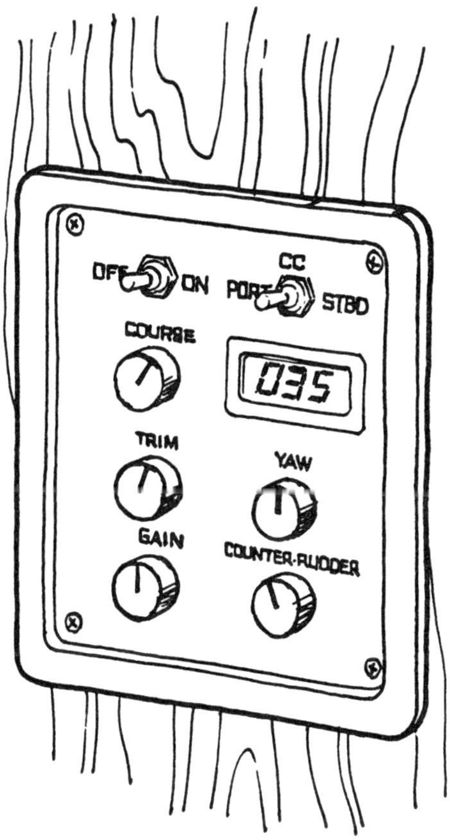

Permanently installed autopilots often have unobtrusive control panels you can mount flush with a bulkhead or dash. Although this unit has about all the adjustments you could want in a good unit, the functions of trim, counter rudder, and gain are now being taken over by microprocessors.

unit. On most proportional models the rudder has a potentiometer attached to it by a small tiller and tie-rod; the potentiometer feeds higher or lower voltage to the control box depending on whether the rudder is hard over or nearly centered. Some of the heavily electronicized systems eliminate the potentiometer with its corrosion-vulnerable metal contacts by simulating rudder angle in a control unit circuit. Such a system senses the drive unit's motor action, specifically the strength and polarity of the signals fed into the drive unit. Comparing these values to that assigned to a neutral helm, the circuit determines rudder angle. Although this system is more reliable than a mechanical system, it may not be quite as accurate.

Counter rudder, a function increasingly being automated, is tied to the amount or speed—or both—of rudder response. Counter rudder keeps a pilot from overshooting the course while making a correction.

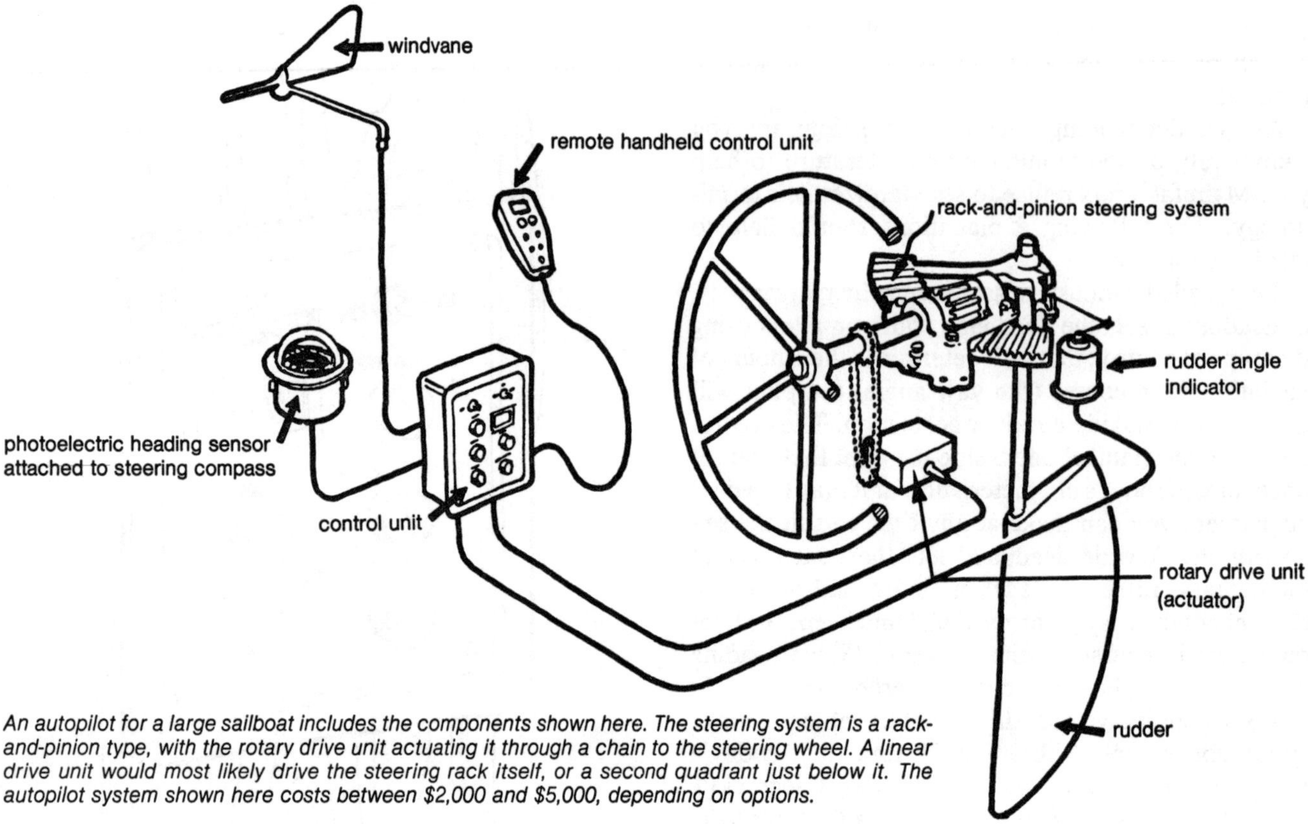

An autopilot for a large sailboat includes the components shown here. The steering system is a rack-and-pinion type, with the rotary drive unit actuating it through a chain to the steering wheel. A linear drive unit would most likely drive the steering rack itself, or a second quadrant just below it. The autopilot system shown here costs between $2,000 and $5,000, depending on options.

Some pilots have an adjustable counter rudder function to cope with various sea and load conditions. On others you set the counter rudder once for your boat's characteristics. The most recent microprocessor models have automated counter rudder to be proportional to the error, just as the rudder response is. In any case, the pilot reverses the drive motor long enough to stop the boat crossing the course line when it is returning from its off-course angle.

Trim (bias, offset) compensates for a persistent heading error, such as from extreme weather helm, only one of two engines working, or a strong, steady wind blowing the bow to leeward. Some pilots have a *trim dial*, which orders the rudder off center to keep the boat on course. This is a manual adjustment you make based on visual references or on experience. You can't feel the tiller or wheel for weather helm, because the pilot has taken over. Sophisticated pilots may have an automatic *trim circuit* that deals with accumulated error by measuring the total error developed over a period of time, averaging it to see whether the boat has deviated more to port or to starboard overall, and then applying the rudder to counteract the bias. It may be a periodic response or a steadily averaged and corrected one.

Trim corrections, however, cannot deal with all the effects of current- and wind-induced leeway. The boat may keep to its heading yet still drift off track. To cope with these influences, an outside position-finding device such as a loran or SatNav must be incorporated into the system. Many permanently mounted autopilots have a loran interface (National Marine Electronics Association 0180 or one of its derivatives), and almost any loran on the market can be plugged into the pilot. The loran's continual position updates are compared to the chosen course, and the pilot is directed to a heading that puts the boat back on track.

Whatever the mysterious workings of the control unit, the result is an amplified signal that commands the drive motor that actuates the rudder. The actuator may turn the steering wheel or tiller in the case of a cockpit-mount pilot; the rudder cables, quadrant, or steering rack in a belowdeck system; or a hydraulic pump in a hydraulic steering system.

To adapt to the myriad installations, there are three basic actuator configurations: linear, rotary, and hydraulic. Manufacturers offer one, two, or all three types, depending on the extent of their product line. Probably all tiller-activating pilots have linear drives,

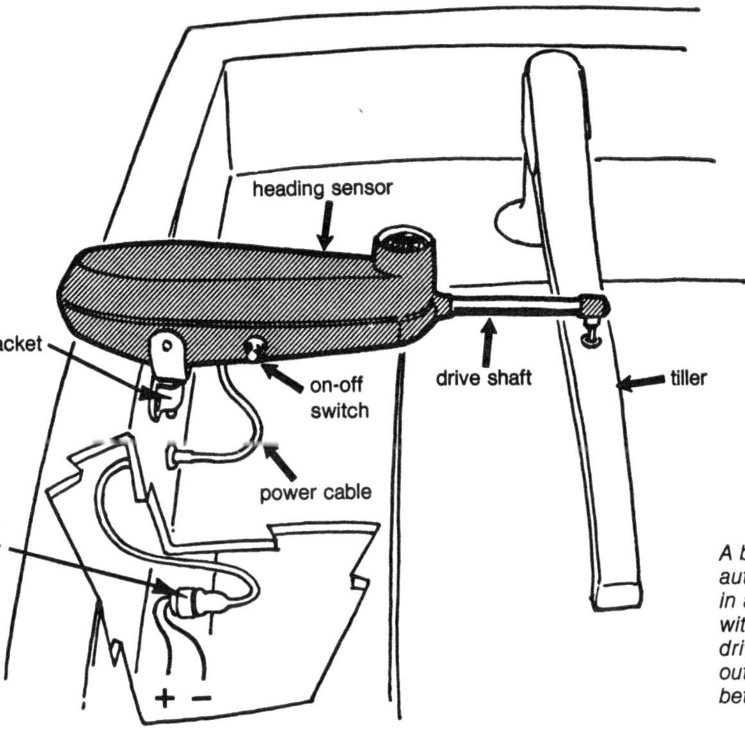

A beautifully simple setup is a tiller-driving autopilot that can be rigged or taken below in a minute or less. Some models come with the heading sensor separate from the drive unit, so it can be mounted below out of the weather. Most portables range between $500 and $800.

in which a servomotor (a small, high-torque motor often used to move aircraft control surfaces) turns a long, threaded shaft. Riding the shaft is a collar to which the tiller shaft is attached. Linear-drive units for belowdecks are no different and usually are attached to a small auxiliary tiller fitted to the rudderpost.

Rotary-drive actuators are offered by more manufacturers. They are the type most used on powerboats and on many large sailboats. They can usually be mounted in more places than can the linear type, since they don't have the long drive shaft to accommodate, and they needn't be positioned next to the rudderpost. Installation may be more complicated, though, requiring a chain to be spliced into the steering cables, a new sheave or two, and perhaps more structural modifications for mounting the new hardware.

Least common in sailboat autopilot systems, though often the most sensible choice for powerboats, is the hydraulic actuator, in which the motor drives a pump tied into the boat's steering lines. Few sailboats have hydraulic steering due to the lack of feel inherent in the system. In a powerboat with anywhere from two to four steering stations (pilothouse, flying bridge, tuna tower, and cockpit in a big sportfisherman), hydraulics make for easy and reliable hookups. There are plenty of hydraulic actuators available, since hydraulics are common aboard commercial fishing and workboats.

How much drive power will you need? Drive units are cataloged along rough guidelines according to boat length, but hull shape, displacement, rudder efficiency, and helm balancing ability, as well as the likely cruising conditions, all play a part. To get a good idea of a sailboat's requirements, sail the boat in the strongest conditions in which you are likely to use the pilot, and at the point of sail (usually a close reach) that develops the most pressure on the rudder. You can simulate heavier conditions by overtrimming sails and creating maximum possible weather helm.

For wheel steering boats, either power or sail, you can then measure the approximate force needed to turn the rudder with a formula suggested by the Benmar Company: Hook a fish scale to a spoke at the wheel rim, and pull enough on it to just turn the wheel, keeping the scale tangential to the wheel. On the powerboat, try the procedure at different throttle speeds. Higher speed turns should yield higher helm loads. Multiply the reading on the scale by the wheel radius. Multiply again by the number of turns *lock to lock*. The result is helm torque in inch-pounds. This figure can be compared to torque figures on the spec sheets (to compare with manufacturers' figures measured in foot-pounds, divide the inch-pound figure by 12).

For a tiller pilot, simply hook the fish scale to the tiller where you expect the pilot to be mounted. Measure the force in pounds needed to move the tiller. Compare this number with the power ratings in the spec sheets.

As important to the sailboater as rudder turning power is power drain. How many amperes will the pilot suck out of your batteries while steering the boat? The answer varies, once again according to boat design, pilot installation, and sea conditions. While on standby, the pilot should draw a negligible amount of power, just enough to sense heading or wind direction. A typical draw for a pilot on course and not driving the actuator is 0.10 amp. Under "normal cruise conditions," which few manufacturers define, tiller-drive pilots draw from 0.20 amp to 0.33 amp. Peak loads might call for three to five times as much power, but in such conditions you will be shortening sail and retrimming to reduce the steering effort, anyway. Permanently installed pilots list a wide range of power drain figures, from as little as that for the cockpit-mount models to as much as 10 amps or more on "cruise." Some drive units are rated much higher, but the rudder may well be hard over and stalled before the limit is reached. In any case, quoted power drain figures are of value in comparing spec sheets only when you know the numbers were obtained in the same way. How much power the system will use on your boat is best estimated by the dealer or manufacturer, who has installed his pilot on similar boats and tested it with an ammeter. Be sure you know your batteries' capacity when you talk to the dealer. All things considered, a sailboat will do better with a system that draws a modest amount of power and can be overwhelmed by the rig in high winds. When that happens, it's a sure signal to shorten sail and balance the helm. This system is more economical, and it's safer.

INSTALLATION

Installing the pilot may involve more work than you think, especially if you have a mechanical steering system in place already. The simplest mechanical tie-in would be a rotary drive actuator turning a chain or gear on your boat's steering wheel shaft. Two reservations: There isn't often room under the dash or bridgedeck of a powerboat for a big drive unit, and the wheel would need a clutch to disengage it when the autopilot is in command. Otherwise, the wheel could be a dangerous and unstoppable moving object should someone's fingers get caught between the spokes.

A more complicated and common mechanical system would see the drive unit next to the rudderpost, driving the steering quadrant directly via linear drive shaft and tiller, or through gears or chain via rotary drive. This system has the advantage of providing a complete backup steering system in case the manual system of cables and sheaves breaks. The layout of the boat's structure, the position of the rudderpost, its angle (vertical, or raked forward or aft, as is sometimes the case in traditional boats), and the type of mechanical components used (cable and sheaves, pull-pull flexible cables, rack-and-pinion gear, or even worm-gear steering) determine whether a given drive unit will fit. This is where a professional is valuable. He has seen many installations, knows what various manufacturers have to offer, and can help you select the one that delivers enough power and fits the space with a minimum of parts and labor.

Hydraulic installations are the easiest. The drive unit (a pump, in this case) can go almost anywhere that's handy, since you can run the hydraulic tubes from it to their junction with the main line. The hydraulic system can be designed as a self-contained backup to the manual steering, too. Install it as a separate system with its own lines and a ram on the rudderpost. It's more expensive that way, but it does give you redundancy in a critical area. If you have a hard time fitting mechanical autopilot components to a manual system, it may even be possible to fit a hydraulic drive unit and associated fittings. Then you have your helm feel through the manual steering, and hydraulics for the autopilot. This is the system installed by the president of a major autopilot company on his own sailboat, so it has the support of at least one expert.

Your boat may well need structural reinforcement where the drive unit is to be located, as well as a mounting stand of whatever configuration is in order. Be prepared also to modify your existing steering system to accommodate the new components. You may have to reroute cables so the pilot drive can be spliced into them. You may need a new steering quadrant that can carry both manual and autopilot cables. Remember again that a clutch should be

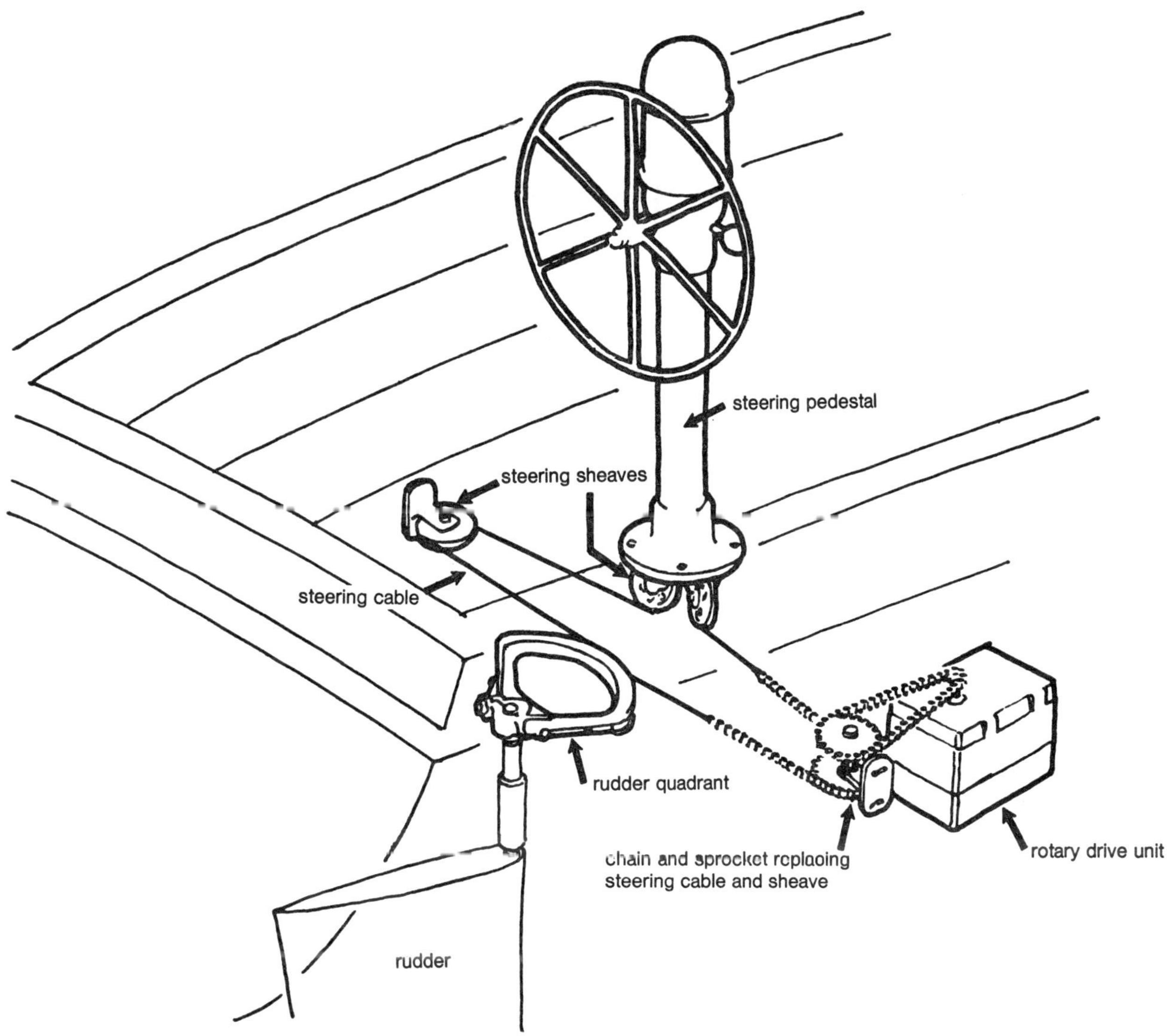

A common system for pedestal steering includes a rotary drive and a chain substituted for a section of one cable.

installed on the manual system to deactivate the steering wheel, unless it's a part of the autopilot system, as with belt-drive pilots that hook up right to the wheel.

Tuning a belowdecks-mounted autopilot involves going out and putting the pilot on successive straight courses north, south, east, and west, while the compass is checked, the pilot's controls are tried, and various other electrical and electronic systems aboard are turned on and off. You want to detect any interference and isolate the cause. Eliminating it may entail noise-suppression procedures or rerouting the pilot's wiring. You may also want to put an ammeter on the drive unit to make sure it's not working too hard—there may be a bind in the system you haven't noticed at the dock. Here, again, the professional installer can save you time, headaches, and money by checking over the system before you go out and by doing a systematic, efficient tuning job underway.

Autopilots are a real boon to mariners who can't or shouldn't have to stay behind the wheel the entire time at sea. Just remember that the best autopilot is no match for the human hand on the helm when judgments have to be made. The autopilot can follow orders just fine, but it can't see for you or think for you. You're the captain, and you're responsible.

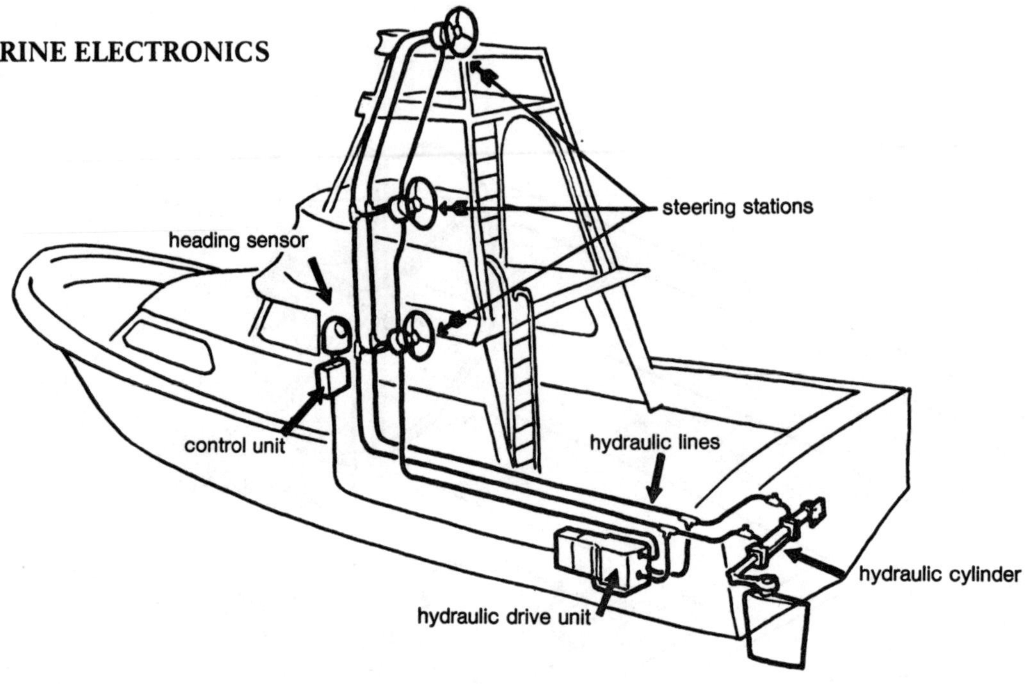

Hydraulic autopilot installations are quite flexible. The drive unit need not be at the rudderpost. Clutches disengage steering wheels from the system. Hydraulic pilot gear runs $2,500 and up.

BUYING TIPS

Choosing an autopilot involves many considerations, and it's difficult to generalize. Every boat is different—if not in hull form and rig, then at least in the sails carried or the engines installed; in the interior layout and the amount of gear and provisions stowed below (which can change the displacement and trim considerably); or (on a sailboat) in the battery's storage and charging capacity. Not least important is the intended use of the boat: Must it sail upwind as well as downwind under the pilot? Will it be raced under the pilot? Will it be used for low-speed trolling in large seas, or fast motoring port to port across a choppy Great Lake? All these factors will have a very definite influence on your selection.

Keep it simple. If you can mount a powerful enough portable model in your sailboat's cockpit, by all means choose that over a permanent belowdecks system. The price differential is one reason, but the complexity and weight of the permanent gear should confine it to boats needing high-power systems. If you can go with a portable unit, you can safely buy it through the mail from a discount house. The installation is simple, the tuning straightforward. Any problems are resolved simply by returning the unit to the seller or manufacturer through the mail.

If your boat—power or sail—is big enough for a permanent system, you would be well-advised to buy from a local dealer. What comes in the box is only a portion of the overall cost in time and money of a particular installation. Unless you are an electrical engineer and are willing to spend a good deal of your summer working out the bugs, you won't save money buying discount through the mail.

Other tips: Don't choose a borderline drive unit. If your powerboat appears to fall between two sizes of drive unit, opt for the larger, more powerful one unless a professional can show you why the smaller one is adequate. An autopilot is not worth the weight and complication if it can't stand up to the job. Likewise, the larger a boat is, the more precise its autopilot ought to be. It takes a lot of energy to make a big boat change course, and the finer the adjustments needed, the less needless wear and tear on the unit. Furthermore, a big boat off course is a dangerous juggernaut.

The autopilot industry is undergoing a major transformation, in which newly designed and engineered microchip models are replacing the older, electromechanical, manually adjusted models. The results are better efficiency, lower power drain, more features, lower prices, and eventually better reliability. Eventually. Until the newer models have been around long enough to have all the bugs shaken out—say, three years—you may save yourself some worry by sticking to a well-known, well-proven model, and leaving the testing to the solo sailboat racers.

9 Radar

by Gordon West

At night or in the fog, marine radar will show you where you are, where you have been, and where you are going. It will let you spot surrounding dangers—buoys to starboard, the jetty to port, or other boats around you in the thick of the night or the height of the storm. More radars have been installed by sail and powerboat owners during the past few years than ever before, convincing evidence that the advantages of radar for piloting and safety are now being recognized by an increasing number of mariners.

For years, radar (an acronym that stands for RAdio Detection And Rangefinding) was considered an expensive proposition—cumbersome to install, heavy on the electrical system, and difficult to use. In the last five years, however, thanks to foreign manufacturers, radars have made a quantum jump in technology (and popularity) with price levels remaining almost the same. Quite simply, you now get almost twice as much radar for the buck as you did five years ago, and your boat's electrical system will love it—the new sets draw about half the power of the older ones.

HOW IT WORKS AND WHAT'S AVAILABLE

Marine radars work on essentially the same principle as echo sounding. Radar waves are timed as they leave the antenna unit, bounce off targets, and then re-enter the antenna system, where they are fed into an amplifier and then the display unit. Unlike the depthsounder, which points only downward, radar signals are beamed out steadily, sweeping a continuous circle and thus giving direction as well as range. Since radar radio waves are transmitted on frequencies much higher than our marine VHF band, the signals travel line of sight (with a 20 percent refraction over the horizon). At such a high frequency, radar waves are very reflective, as they need to be.

Inside the radar is a timing device that measures time delays in microseconds (millionths of a second). At close to the speed of light, a radar pulse will travel 328 yards in one microsecond. If one microsecond passes from the time a pulse is transmitted to the time its echo is received, that burst of energy has traveled 328 yards—164 yards out to the target and 164 yards back. The radar displays the target distance as 164 yards after one second of internal calculations. Antenna position information is also relayed to the display unit, so that not only the range but the precise bearing of the target is displayed on the scope. Finally, the signal strength of the target on the scope is an indication of its size and density.

The basic components are the antenna unit, the transmitter and receiver circuitry built into the antenna unit, and the display unit, which contains the timing and computer circuits, the cathode ray tube, and the power supply. The transmitter, receiver, and antenna system are housed in a 30- to 60-pound package, and the display unit is a lightweight 20-pound package that mounts below. A one-inch-diameter cable connects the antenna unit with the display unit, and the display unit gets its power from a relatively small pair of 12-volt wires.

Inside the antenna unit, the transmitter section develops powerful energy pulses on a frequency near 9,400 MHz. The length of each pulse of energy might be as short as 0.05 microsecond for short-range calculations or as long as 1.0 microsecond for long-range calculations. Shorter pulses at closer range give better target definition and resolution. On the short-range scale, the pulse repetition rate is also increased, to as high as 2,000 to 2,500 pulses per second. This allows the receiver to listen for echoes between each transmitted pulse. At longer ranges the echoes will take longer to come back, so the radar will automatically slow down to approximately 800 to 1,200 pulses per second with pulse lengths many times longer. Again, the radar must immediately switch to receive after each high-powered pulse is transmitted in order to allow the receiver section to pick up the echo. A modern, microprocessor-based radar automatically selects the optimum pulse repetition rate and pulse length depending on which range you have selected.

Most modern radars offer a minimum of five selectable ranges, the shortest of which may be ¼ mile, and the longest, 16 to 20 miles. There *are* radars with more power and longer range, but long range is the *least* important feature for a small-boat radar system. Since radar waves don't bend over the horizon, it makes no sense to buy a 40-mile radar when you can only mount it high enough to put the horizon 8 miles distant. If you are planning to operate along coasts with high mountains close by the shore, you might choose one of the longer-range radars. Remember, though, that for a radar to detect a target at a maximum range of 20 miles, your radar antenna must be 10 feet or more above the water and the landmass must be more than 170 feet above sea level.

Small-boat radars normally transmit with about 3,000 watts of peak power in every pulse, while the receiver section is capable of detecting signal echoes as low as one-tenth of a microvolt in intensity. The antenna completes 30 to 40 360-degree rotations per minute.

The antenna unit may be encapsulated under a fiberglass radome, or it may be out in the open. Choose the former if you are going to be mounting it aboard a sailboat halfway up a mast, since sails and rigging would foul an unenclosed antenna. In either case the assembly consists of a horizontal reflector that beams out and retrieves the radar pulses and echoes. Beam widths may be from 1½ to 5 degrees depending upon the width (or span) of the antenna. Open antenna reflector units are commonly 3 feet long, but a longer unit (4 feet, for example) will concentrate radar energy in a tighter horizontal beam, giving you greater target definition. The small antenna assemblies enclosed in radomes, while they are very compact, might show two fishing boats close together at a range of 5 miles as a single echo on the scope, while the slightly longer, open antenna would show the vessels as two separate targets. Longer antennas lead to better target definition.

The antenna assemblies are designed to offer reliable operation even when your vessel is rolling, the vertical beam width of most radar signals being greater than 25 degrees. In heavy seas, however, a distant target might "disappear" in the trough of a wave at the same time your radar sweeps that sector. Intermittent targets in heavy seas are probably small boats a few miles away.

You may wish to mount the antenna unit as high as practicable to increase the maximum realizable range, but remember that while it might be nice to have the antenna halfway up a sailboat's mizzenmast, it's not really necessary for good coastal navigation and short-range ship detection. Antennas just 10 feet off the water will give you a good view of breakwaters, ships, and any other targets within 4 or 5 miles, and it's what is happening in your immediate neighborhood that concerns you most.

When an antenna is mounted down low, there is always some concern over the potential harm from radar waves. Since no type of radiation can be beneficial to the body, it would certainly be best not to have the radar setup just a short distance away and sweeping by at eye level. Many installations on powerboats do put the antenna assembly directly in front of the operator on the flying bridge, and there have been no reported adverse effects. Probably the intermittent sweeping of radar waves is no more harmful than standing beneath a large commercial broadcast station antenna tower setup. The sensible approach is not to court trouble: There's enough radiation in the slotted antenna reflector to heat up a bologna sandwich. Mount the antenna assembly well above all occupants on the vessel and you can put your mind at rest.

The display unit is the component that has seen some dramatic improvements. At 20 pounds or so, it's quite lightweight, and the internal power supply in the indicator unit is fully transistorized and uses new circuitry in place of heavy transformers. The unit also houses the microprocessor electronics that allow you to select the many different modes of operation for the radar. Most small-boat radars will "see" as close in as 30 yards from the vessel when on their ¼- or ½-mile range scale, and most have five or more selectable scales for short-range, medium-range, and long-range operation. When shopping for a radar, note how many short-range scales each radar offers for detecting targets less than 5 miles away. Again, what is happening close by your vessel in the night or the fog is your primary concern.

A radar set offers other controls besides the range selector, of course. A rotatable cursor allows you to display a straight line that emits from your location in the center of the screen. You can rotate the cursor to put it on line with a target, and the target's relative bearing may then be read where the line intersects the 360-degree azimuth at the scope's perimeter. With this feature, you can plot the bearings of other vessels in your vicinity, and watch their movements. If another vessel draws progressively nearer along the cursor line, you are on a collision course. Some radars also feature audible anticollision alarms that will be triggered when an approaching target closes within a preset range along an imaginary cursor line. Alternatively, an alarm can be set to go off whenever any target, regardless of its relative bearing, is detected within a predetermined range. If you get too close to shore, for example, the alarm will be triggered, calling your attention to the scope.

The electronic variable-range marker is another relatively new refinement in radar display consoles. Instead of having to estimate the distance to a target by interpolating between the fixed range rings that match the range scale you have selected, you can electronically expand or contract the circular range ring to coincide with the distance to a target. The range (to as close as one-tenth of a mile) is displayed digitally.

The traditional radar scope utilizes a cathode ray

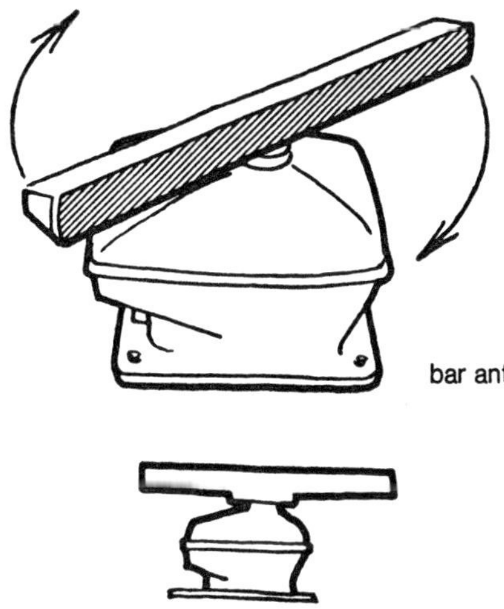

bar antenna

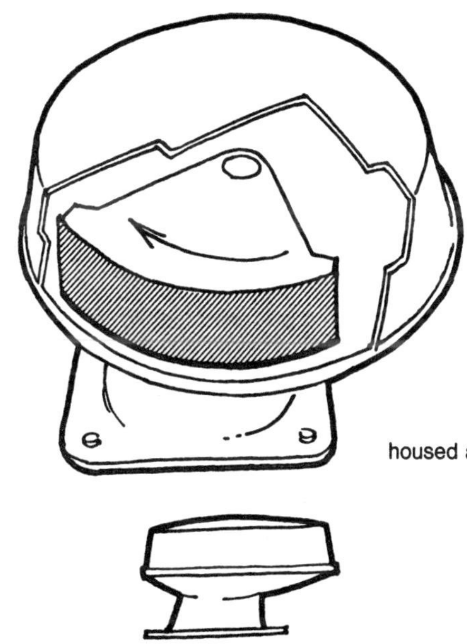

housed antenna

Radar antennas may be of the open type ***(top)****, with a span of 36 inches to as much as 9 feet, or enclosed within a radome* ***(bottom)****. In the latter, the span is seldom more than 30 inches. The antenna systems used on most small boats have the transmitter and receiver section with its antenna rotating motor in a watertight housing directly below the antenna.*

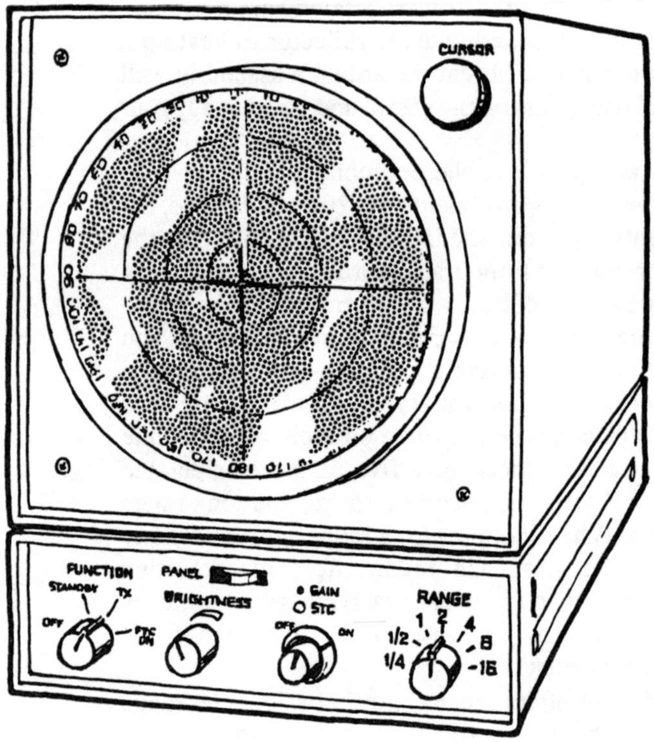

The number of controls on a radar display unit varies with the manufacturer and the radar's maximum range, but every model must have at least four, with one or more serving two purposes. At the upper right corner of this display console is the cursor rotating control for moving the cursor line through the full 360 degrees to obtain the relative or magnetic bearing of a target or headland; at the bottom are the controls to energize the radar and adjust its gain and sensitivity, its range scale, and the brightness of its display. The price range of a new radar (with antenna) is $1,500 to $5,000, and the size of a display unit like the one shown here might be 10 inches, by 10 inches, by 15 inches.

tube with targets and range ring information displayed as bright orange "returns" on the plan position indicator (PPI) scope. Your vessel is located in the center of the picture and emits a faint scanning line that sweeps around the scope every couple of seconds, in synchrony with the antenna rotation. When this scanning line detects incoming signals, it literally paints a picture on the scope showing the bearing and range of the target and its relative intensity on that sweep. The targets are re-illuminated with each rotation of the antenna. Fast-moving targets may appear to leave a phosphorous "wake" as you watch their progress across the screen.

This display system is still used on most radars today. An initial inconvenience is that when you make a tight turn, the shoreline "smears" as old targets decay and new targets appear in different locations. After a while, however, you will get used to this, and you will experience no problem interpreting the radar in heavy seas, tight turns, or simply while plotting fast-moving vessels across the scope.

The only real drawback is that you can only turn up the intensity of the display so far. Most presentations are orange and are easy to see in a reasonably darkened pilothouse. When a cabin admits a great deal of light, however, a hood is required for daytime viewing. It is possible to purchase a hood with a corrective lens built into it to compensate for near- or farsightedness. Magnifiers may be ordered with a radar setup to increase the relative size of the PPI scope presentation. The conventional PPI scope has been around for years and will continue to serve the industry for many years to come.

Recently a new type of display has gained in popularity. Called "raster scan," it looks a little like a video game. The raster scan display gives you instant, digitally processed echoes that may be displayed in sharp contrast to areas on the screen where no echoes are present. You can even choose which color you want the echoes to be. Some models have a white screen with black echoes or a black screen with white echoes. Others offer an orange background with green targets.

Each echo consists of tiny squares, the number of the squares corresponding to the intensity of the echo. Weak echoes may only illuminate two or three squares, while large echoes may illuminate a block of 20. As many as 250 to 500 lines per side of the screen are used to define the squares—a long way from the resolution and sharpness of a television picture, but effective nevertheless. On a conventional phosphorous PPI scope, a gently sweeping shoreline would appear as a smooth curve. On the raster scan, it would resemble a million little bricks comprising a jagged, stepped outline. After a while, however, one can get used to seeing blocks rather than smooth curves.

The echoes on a raster scan display may stay illuminated for several seconds, even though the microprocessor hasn't seen their target return for a couple of sweeps. Then, suddenly, the echo disappears. This, too, you must adapt to if you have previously operated phosphorous-type scopes, in which targets fade gradually.

Some manufacturers have followed in the steps of video depthsounder makers and have assigned different colors to different echo intensities. This may help the mariner identify extremely strong targets that grow in intensity but barely swell in size on the raster scan display.

Probably the biggest benefit of the television-type display over conventional phosphorous PPI scopes is the fact that it is inherently brighter and easier to see in a bright pilothouse. You may never need to use the hood in order to create shade around the scope. Another benefit is that you can add more than one display unit to your system. Once you set the controls down below, you might have a number of additional readouts—each remote readout simply hooking up to 12 volts and a common coaxial cable feedline, just as in cable television. This can't be done with the PPI phosphorous scopes.

It is to be hoped that raster scan, as it continues to evolve, will give us more lines per inch for better resolution. Look for this to happen, and if you're shopping for a raster scan, use the degree of target definition in the display as one of your criteria.

Regardless of the display you choose, you needn't worry about power consumption. At 12 volts, some 20-mile radars pull less than 4 amps. Since the engine is usually running when the radar is in operation, you should never have a problem with the radar draining your batteries.

OPERATING PROCEDURES

Before turning on your radar, doublecheck your marine radiotelephone license and see whether or not it has been endorsed for 9,400 MHz (9.4 GHz) radar operation. If it hasn't, you must fill out a new Form 506 and specifically request radar frequencies from the Federal Communications Commission. As soon as you drop this form into the mail, you can start operating the unit. When your license comes back, you will have the same radio call sign, but with additional frequencies allocated for radar use. You are not allowed to operate radar without the FCC frequency endorsement on your marine radiotelephone license.

Begin using your radar set in the daylight and note how the objects around you appear on the screen. You will notice that radar doesn't see over large ships or around islands, nor will it pick up low-lying rocks at distances greater than 3 to 5 miles. Radar works on a line-of-sight basis.

The type of target will influence its ability to reflect radar signals. Sailboats with sails set are poor radar reflectors. Radar waves simply penetrate sails, and an aluminum mast reflects radar signals in scattered directions. The effect is akin to bouncing a tennis ball off a round pole.

Some targets will make huge presentations on your radar scope. Supertankers turned broadside are an example, as are buoys with radar reflectors. Anything comprised of flat metallic surfaces that intersect at 90 degrees will reflect radar signals with great intensity. This is the benefit of mounting a radar reflector on a vessel.

Since fiberglass is a poor conductor except when wet, fiberglass boats are not good reflectors of radar waves. You will notice, too, that different shorelines give readouts of different intensity. Low, sandy beaches barely show up until you are almost on top of them, while sharp cliffs reflect signals with tremendous intensity. Low, rolling hills give only fair echoes, and sandbars will give almost no return echo at all.

Heavy seas may reflect radar signals, causing confused readings at times. There are controls to help eliminate sea clutter return, however. You can also spot approaching rainstorms with radar, or manipulate the appropriate control to help eliminate rain return.

How sensitive is a radar? A good radar set should pick up a low-flying pelican up to one mile away, or a seagull resting on flotsam one-quarter mile away. The calmer the seas, the higher you can turn up the intensity control for maximum sensitivity and target detection on a phosphorous scope radar.

Once you have gained confidence in detecting targets, the next step is to learn to use radar to avoid groundings and collisions. If you carefully align the cursor over a target, you can watch its progress and ensure that it is not on a collision course with you.

You can safely navigate within a harbor using the radar on its shortest range setting. Enter the harbor a few times in the daylight with the radar set on, and you should be able to rely almost completely on radar for a nighttime approach. Consult your chart for the buoys that appear as targets on your scope as you approach the harbor. Watch the shorelines or breakwaters as you pass through the entrance, and then look for small craft that might have been "hidden" behind a breakwater. As your skill increases, you'll be able to spot and negotiate the appropriate channel in a crowded anchorage, coming within 30 yards of your dock on the basis of radar alone.

At sea, you can use a radar for position finding by measurement of distance and bearing. Spot large mountains on your charts, and locate these landmarks on your display, with the radar set to one of the longer ranges. Distant peaks will appear as small traces toward the scope perimeter. Use the cursor to develop lines of position.

By extension, radar can also be used to determine speed over bottom or vessel drift. It will even tell you when other vessels in the area are using radar, since curved interference traces will appear on your radar with every antenna revolution. Some mariners may curse at these interference traces, but they are really quite helpful. They can, however, be eliminated with special circuitry.

The United States Coast Guard has made it a bit easier to identify certain targets by equipping them with racon devices. Assume you are entering a harbor and need to ascertain which target is the main channel buoy; you'll be in luck if that buoy carries a racon device. It will paint a coded signal on your radar set, removing all guesswork. In the manner of the air traffic controller "squawk" signals that identify aircraft, racon buoys and shoreside racon installations mark the exact spot of a reflected signal. You can expect to see more of these devices in the near future, as radars become more popular.

ADDITIONAL BUYING AND INSTALLATION TIPS

You must really see a radar set to judge it, and for that reason, particularly if you do not plan to install the unit yourself, you would be wise to purchase it from a specialty marine electronics dealer, one that is an authorized sales and service agent for that particular unit. You'll want someone local in case the radar should require service, so that a technician can be brought aboard for the work. Most of the electronics are in the antenna unit, which is not easy to dismount for repair off the boat.

Radar sets require high voltages, and high-voltage circuitry is inherently more prone to malfunction. This is another reason why most radar sales take place at the dealer level rather than through mail-order sources. It takes a skilled technician at a hefty hourly rate to work on your radar, but if you develop a service contract with the marine electronics dealer that sells you the equipment, you can save yourself some money down the road. Most radars require service at least every two years due to their high operating voltages as well as the sensitivity of the magnetron transmitter output tube. If you use your unit a lot, it will require annual peaking and tweaking. For best results, this can only be done right on your boat by an experienced technician. Buying from a specialty dealer—and insisting on a firm service contract—is the surest way to a long and happy relationship between you and your radar.

If you *do* decide to install your own radar, follow carefully the instructions in the manual that comes with the unit. Unless the manufacturer includes an FCC certificate in the packing case, it will be necessary to have a technician "sign off" your installation and make an entry in your radiotelephone and radar logbook as required by the FCC. If you buy the unit through a mail-order house, be prepared for this added, not-insignificant expense. The local marine electronics specialty dealer from whom you *didn't* buy the unit may not be eager to send down a technician to sign you off for just a few bucks. Indeed, he may charge you several hours of service time for this sign-off procedure.

A few marine electronics mail-order companies will certify a radar before it is shipped to you. If this can be done, it will save you some hard-earned money once everything is aboard and operating. Since radars are so sophisticated and self-adjusting, a do-it-yourself installation is feasible *if* you are good with tools and can devise an appropriate antenna-system mount. Some manufacturers offer pedestals for mast-mounted antennas, and there are countless other possible approaches. Remember, though, if you're going to mount the antenna high above the deck, that there may be tremendous stresses on the installation when your vessel is pitching in heavy seas. Since the antenna unit contains almost two-thirds of the radar's electronics, make sure it's bolted down firmly. Replacing it would be expensive.

The antenna assemblies are matched for the radar set you are buying. When you buy the radar, you buy it as a complete setup, so you needn't worry about matching the antenna assembly to the radar. If you're given a choice of antenna lengths, try to accommodate the slightly longer antenna for better target definition. The cable connections between antenna and display unit are straightforward; you can hook it up yourself simply by following the directions and screwing in the wires by the numbers.

Prices for radar setups have been holding steady for the past two or three years at $2,000 to $3,000. Don't look for any dramatic price drops in the near future. Thanks to foreign competition, however, more and more features are being made available for the money. This trend is likely to continue.

MAINTENANCE

Keep the display unit out of the weather to protect delicate circuitry within the console. Don't leave it in a damp flying bridge for any length of time. If you're going to be storing the boat for some time, take the display unit home to store in a dry place. Never seal it in a plastic bag—that only collects moisture.

Now and then, give your antenna radome a good bath with warm, soapy water. Radar waves have a hard time sounding through accumulated soot. Never paint the antenna dome—lead-based paints, especially, will literally soak up radar signals. The painting of fiberglass radar antenna domes is a common practice, but a very poor one.

Never tune the intensity up too bright. It should be just bright enough for viewing. The phosphorous cathode ray tube displays, especially, are easily damaged by a too-high brilliance setting.

10

Performance Instruments and Integrated Systems

by Freeman Pittman

As with depthsounders and some of the navigational gear described elsewhere in this book, instruments—particularly sailing performance electronics—are no longer a case of one function–one dial. The range of capabilities available from a variety of equipment is so great as to be impossible for most of us to master. But there is a cardinal rule to follow while wandering through the wonderful world of time, speed, distance, heading, and angle: Don't buy it if you don't need it.

You may think you need it now; you may think you'll grow into the system as you get more comfortable with it. But remember the computer in your office or home. Are you using it as much as you thought you would be when you bought it? Are you using it for only a few functions, without exploring its potential? It's the same with the sailing computer systems and many of the more basic electronics we take for granted on so many boats.

A 24-foot boat I frequently sail on had a digital speedometer and log, a depthsounder, and a VHF radio ordered for it. After two seasons on the water, not one of these basic items has been taken out of its box. The owner hasn't gone to the trouble of drilling holes in the boat bottom for the depth and speed transducers (or hasn't gotten the nerve yet), and the VHF antenna and cable apparently aren't worth the trouble of installing, either. Last summer the boat didn't even have its battery connected for the navigation lights. Which was fine, because it didn't go out after dark. Nor did it ever venture out of protected, well-chartered waters or go out on days when we couldn't see exactly where we were. That boat doesn't need electronics, although many of its sister ships are so equipped. It's better off without the holes in the hull or the additional complications electronics are bound to add. I predict the never-installed electronics will soon be sold.

It's knowing what you will do with your boat that determines your electronic requirements. With that in mind, let's look at what you can buy and what it can do for you, starting with the single-function units.

SPEEDOMETERS FOR POWER OR SAILBOATS

The most basic indicator, along with a depthsounder (see Chapter 2), is the speedometer or log unit. Of the two functions it displays, speed and distance, the latter

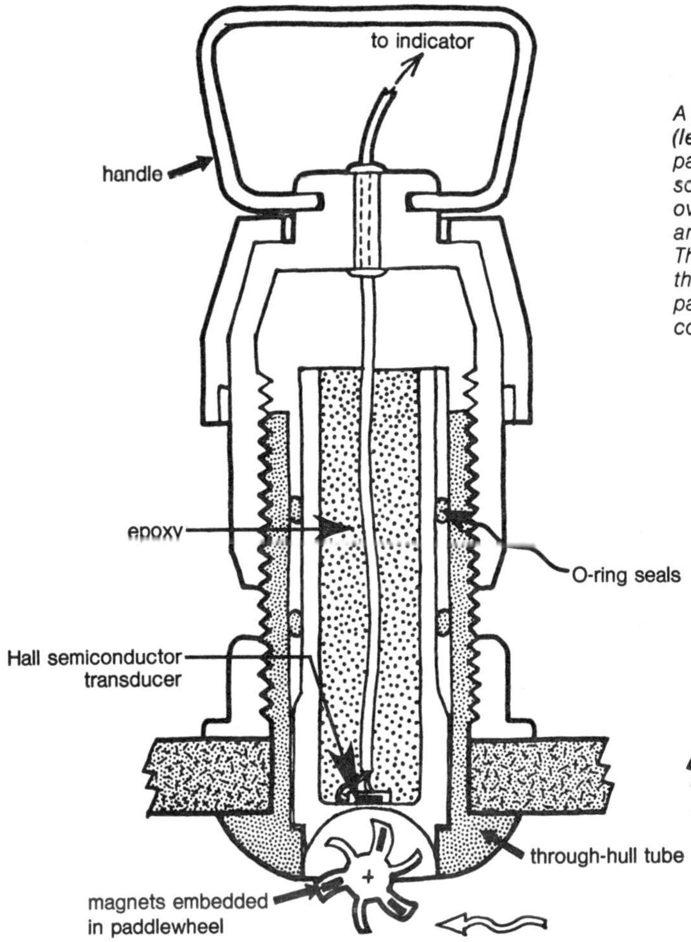

*A paddlewheel boat speed transducer in its through-hull tube **(left)** and a flush-mounted through hull **(right)**. Mounting the paddlewheel at the end of a protruding through hull can lead to some distortion of water flow, but the instrument's calibration will overcome any inaccuracies. Most through-hull units nowadays are plastic and must be mounted with compatible adhesives. They are also more vulnerable to damage than bronze tubes, but they present a small target. The boat speed indicator, paddlewheel, and through hull plus 50 feet or so of cable can cost as little as $200 or as much as $800.*

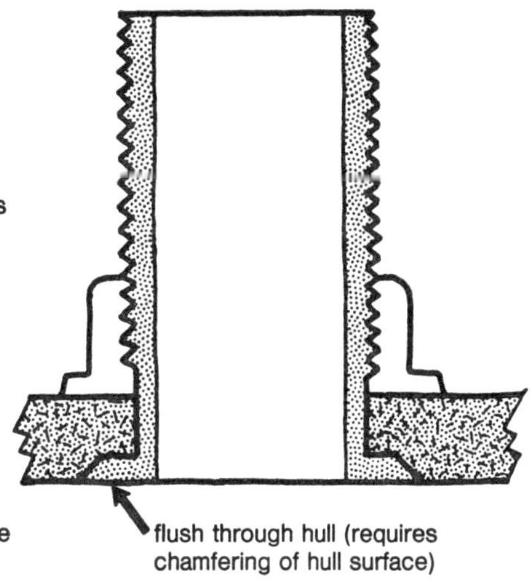

is the more important. You can convert speed to distance and vice versa, but for navigational purposes, knowing how far you've traveled is essential to knowing where you are or ought to be. How fast you were going along the way is incidental. The guts of the unit may vary according to manufacturer, but almost all modern speedos share certain features. The through-hull unit that generates the electrical signals used to calculate speed is normally a small paddle-wheel impeller in a cylindrical housing. A shielded cable takes the signal to the display unit, where it is amplified, timed against a crystal oscillator, and displayed on a digital liquid crystal display (LCD) or on an analog, pointer-type dial. The log portion of the display unit may be a mechanical readout like your car's odometer, or it may be a digital display that alternates with the speed readout whenever you flick the switch.

Though most are similar in form and operation, transducers are the weak link in the system. There are a number of reasons why they can't be as accurate as they would have to be to keep up with the other electronics aboard, and there are several ways that they can fail altogether. Many are the solutions that have been tried, but right now, the vast majority of speed logs employ the reasonably trustworthy and surely simple paddlewheel.

The plastic paddlewheel rotates on an athwartships axle. Molded into the paddlewheel, or glued to it, are magnets, and inside the through-hull unit, just above the impeller, is an electrical reed switch or Hall-effect semiconductor carrying a small but constant voltage. As the magnets in the impeller pass by this sensor, they generate a pulse that goes to the indicating unit for processing. So many pulses, say 50,000, count as a nautical mile. There is your distance. Factor in the time element and you get speed. A few manufacturers use the impeller as a tiny generator armature, which develops enough power to drive the speedometer without battery input except for night lighting.

A typical indicator for boat speed and distance. Dimensions: 5 inches diameter over the bezel; 4 inches over the barrel, which protrudes through the bulkhead about 3 inches.

The trouble is, the paddlewheel doesn't always rotate at the speed it should to give you the correct reading for speed through the water. In the case of a sailboat beating to windward, the boat's leeway combines with the asymmetric underwater shape of the heeling hull to cause water to flow over the impeller at an angle. This crossflow makes for slower impeller rotation than if the boat were sailing upright without leeway. If the impeller is too close to the keel's leading edge, on the other hand, water flowing around the keel can actually accelerate the flow over the transducer. Then there is heaving and pitching of the hull, which make an impeller mounted too far forward vulnerable to breaking out of the water on wave crests. If the impeller is mounted in any spot on the hull that experiences frequent water turbulence, accuracy must suffer. And if it is on one side of the hull rather than on the centerline, it will probably read differently on one tack than on the other. Quite often this is the case, because the hull can be too thick at the centerline for a through-hull installation. The obvious if somewhat expensive solution is to mount an impeller on either side of the hull with a gravity switch to change automatically from one side to the other. In any case, the through-hull unit should always be mounted forward of amidships, preferably away from the keel or underwater fittings, so it will be in smooth water.

Displacement powerboat hulls needn't worry about heeling, tacking, or leeway error, but they, too, need clean water flow over the transducer. Planing hulls will need the transducer mounted well aft, since the bow sections will see a lot of spray, or even daylight from time to time.

One other factor often brought up when discussing the paddlewheel or other impeller types is the "boundary layer," the water closest to the hull. Due to the hull's surface friction, the boundary layer tends to move forward with the boat as it passes through the medium, so an impeller mounted in this layer will read low. How far from the hull the boundary layer extends depends on several factors, but if your speedometer impeller is mounted correctly, it should protrude far enough into clean water to be accurate through most of the boat's speed range. At speeds over perhaps 8 knots, though, the boundary layer thickness may build up enough to reduce speed readings by as much as 5 percent.

Manufacturers recognize the aforementioned problems and have ways to deal with most of them. Most speedometers have a calibration knob on the back that allows you to change the number of impeller rotations needed to count off a given distance. You run a measured course—from one buoy to another, for instance—in deep water with a steady tide or current. Motor the boat at a constant speed, judging as best you can what your typical cruising speed would be. Mark down the exact distance logged electronically between the buoys, then turn and motor back on the reciprocal course. Note the total distance and divide by 2 for the average as your log sees it. Then find the actual distance from your chart and compare. The difference can be dialed into the instrument on the calibration knob according to the manufacturer's instructions. Errors of 5 percent or more can be eliminated in this way.

What can't be adjusted for is an impeller fouled by marine growth or floating debris. Check your impeller occasionally, and clean it as necessary. Some units are painted with antifouling paint at the factory. This can be renewed, but be careful not to let it gum up the works.

For all the potential problems, the paddlewheel and its close cousin the propeller-type impeller work pretty well. They're also relatively economical. But inventive minds are constantly trying to improve on the things that aren't yet perfect. Here are some examples.

An alternative that seems most common in England is the electromagnetic (EM) transducer, with no moving parts. It has two little terminals about half an inch apart, barely protruding from the hull. Between them flows a small current, induced by a magnet in the through-hull. Water flowing through the current distorts it, and the faster the flow, the greater the distortion. The difference in voltage from the steady state current is measured and displayed. The concept is simple, and the installation is clean and smooth, which sailboat racers in particular like. But slime and other growth can affect this transducer more easily than the paddlewheel types, and the boundary layer is indeed a factor with the EM transducer, since the terminals are no more than an eighth of an inch high. That means that after calibration, the unit will still be more accurate at some speeds than others.

The force of water on a small, protruding rod can also be translated into boat speed readings. The rod—or wand, as it is sometimes called—is deflected by the water pressure, which causes its electrical resistance to change. These variations are transmitted through the cable to the indicating unit, different values corresponding to various speeds. Such a transducer is better suited to powerboats than sailboats, since its low-speed accuracy is not too great, while its ruggedness serves well at high speeds. Nonetheless, few are in use today.

The nonrotating system most likely to succeed as a viable alternative to the impeller transducer is that introduced for pleasureboats by Brookes & Gatehouse and widely used in England. Developed as a result of submarine acoustics research, the system consists of two small sonic transducers, similar to miniature depthsounder probes. They are mounted three feet apart on the hull or on the hull and keel, facing each other along a fore-and-aft axis. There must be clear water between them, a condition easily satisfied on most of today's sailboats, which feature a cutaway forefoot. On many powerboats the units can be fitted to hull and skeg. Each transducer sends out a sound signal. The signal sent aft is hurried on its way by the water flowing with it, and the one sent forward is slowed. The speedometer's microprocessor subtracts the slower signal speed from the faster and comes up with twice the boat's speed, which it then halves and displays. The device does away with boundary layer problems, moving parts, and opportunity for clogging. Crossflow around the keel's leading edge must be calibrated for. Fouling from marine growth must still be prevented, and the transducers cannot be removed from the bilge for inspection. Also, turbulence can affect readings, just as for depthsounders, so the transducers must still be sited in clean water. The next couple of years will see this fascinating mechanism tested.

The coaxial cable carrying the transducer's signal can pick up interference from other electronics if it's cut or damaged. Don't cut it to length without first reading the manufacturer's instructions—they may warn you not to cut it at all. Good-quality instruments have the transducer and the transducer end of the cable "potted" or molded in a solid material such as epoxy, from which the cable runs about 10 feet or so to a connector. The connector enables you to remove the through-hull fitting if needed without having to cut or pull out the main run of wire, and removing the connector from the transducer by several feet keeps it out of the bilge. Similarly, there should be a connection at the display unit end to allow it to be removed without disturbing the cable.

WIND SPEED INSTRUMENTS

Wind speed instruments have much in common with speed-logs. Instead of a rotating impeller underwater, they have one aloft, with three little black cups spinning in the wind, or, on at least one masthead unit, with a propeller like an airplane's. The transducer may work with a rotor spinning a magnet past a reed switch or Hall-effect sensor, as in the boat speed unit, or it may be designed, once again, as a tiny generator, developing voltages proportional to the speed of the rotor. Yet another interesting approach puts a crown-like fitting on the shaft, where it spins past an LED and photocell, alternately making and breaking the light beam circuit between them.

The transducer is normally mounted on the end of a stalk facing forward from the masthead. This gets it into clear air and away from antennas, spinnaker halyards, and the chute itself, which might otherwise foul it. You might be surprised by the amount of aerodynamic interference other hardware aloft—including the wind unit's stalk—can develop. The masthead

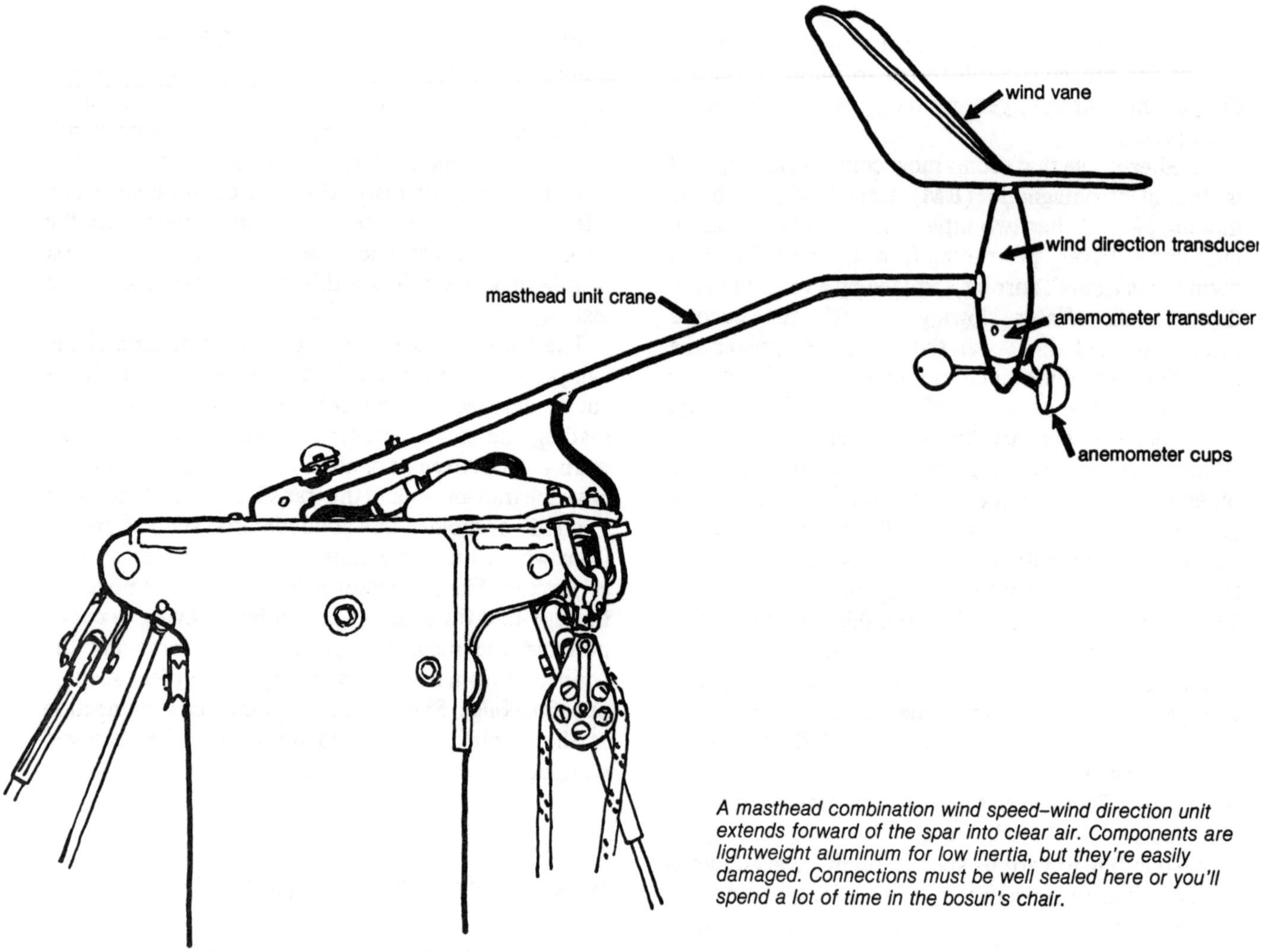

A masthead combination wind speed–wind direction unit extends forward of the spar into clear air. Components are lightweight aluminum for low inertia, but they're easily damaged. Connections must be well sealed here or you'll spend a lot of time in the bosun's chair.

unit must be robust, but with very light moving parts, and it must be water- and corrosionproof.

The masthead installation requires effort and care. Be sure the unit is securely positioned on the fore-and-aft axis and that it indeed extends forward from the mast, unless the manual specifically states that you can mount it extending aft over the mainsail. The windspeed unit doesn't care, but the wind direction indicator that usually comes with it may not operate properly unless facing forward.

The apparent wind direction unit tells you where the wind is coming from relative to the boat's forward progress. It is no more accurate and probably less reliable than a mechanical masthead fly, but you can read it on a dial at deck level instead of looking aloft, possibly into the sun (or the night). The readings can be more precise with the electronic unit, detecting small shifts of less than 5 degrees. And you have to have the unit if you want the kind of computed data obtainable from an integrated instrument system.

The wind vane rotates on a precision bearing, and it should balance perfectly. You can be sure it does by holding the unit on its side and watching to see if either the nose or the tail of the vane drops. Often the nosepiece is adjustable for balancing the vane, or you can add small weights.

The vane's shape varies with the manufacturer's aerodynamic philosophy. Some vanes are long with small surface area, like the tail feathers on an arrow. Other vanes are short but larger. These have the advantage of reacting more quickly than the longer ones, since they usually have less inertia, and they will not over-rotate on the larger shifts. Regardless of shape, the vane should be very light to overcome bearing friction at low windspeeds.

To send electrical signals, the vane shaft may have a

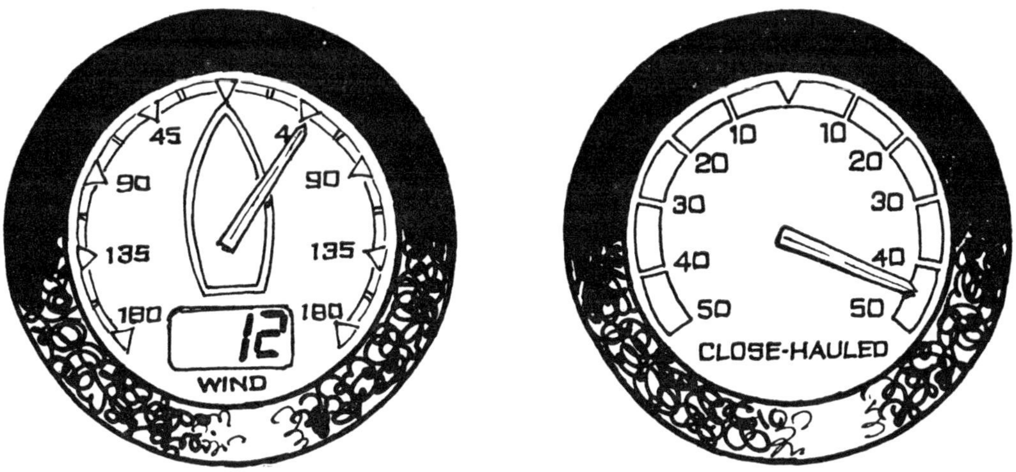

*Standard readouts for apparent wind angle on the 360-degree scale **(left)** and on the expanded close-hauled scale covering only the first 60 degrees on either side of head-to-wind **(right)**. The close-hauled option is helpful to racers. The angle displayed is relative to the boat's heading. Wind speed is often combined in the same instrument, as it is here. The package (including sensors) costs $600 and up.*

metal wiper or two attached to it, and these sweep a potentiometer coil wrapped around the shaft. The varying voltages developed at the masthead are amplified and processed in the display unit and used to drive the pointer on the indicator dial.

The display unit, whether for boat speed, wind, water depth, or piloting data, takes one of several shapes and forms. There are the traditional barrel-shaped models with a flange or bezel for mounting through a bulkhead. Other models are similar but with a square bezel. There are nearly flat units that mount on the bulkhead. These do not require the drilling of any major holes, the only bulkhead holes being for fasteners and wires. There are wheelhouse units for powerboats or pilothouse sailboats, using brackets you can place on the dash ahead of the wheel or hang from the headliner. These are bulky, but they allow you to tilt the unit to avoid glare and to remove it easily for safekeeping. As another alternative, several types of information may be combined in one housing, either on separate displays or sharing a single readout. Unless this kind of instrument is right by the helm where you can change the display easily, opt for separate displays; you certainly don't want to give your navigator the wrong information when he calls for current boat speed—you may be reading wind speed, depth, or water temperature without realizing it!

Do you want analog readouts or digital? The choice is strictly personal. Most of us have gotten used to reading digital displays, and where the transducers can send accurate enough information, digital displays can give more discrete readings. Many speedometers read out in tenths of a knot, and some even in hundredths. That implies much greater precision than the system is capable of delivering, however, given the transducer's limitations. Besides, the speedometer electronically averages the signals over a period of 2, 5, or 10 seconds. If it didn't, your eyes couldn't keep up with the number changes.

Every point made in the Depthsounder chapter about LCD versus LED displays holds true for speedometers and wind instruments. Apart from contrast and clarity, though, pay attention to how wide a viewing angle the display has. LCDs are not noted for showing much unless you're looking at them nearly head on. If you steer from the windward or leeward rail, as do many sailors, the instruments will have to be mounted well ahead of the helm or in one of the swiveling instrument pods you can get for many steering pedestals.

Instrument lighting is something you don't often get to judge in advance, but do try. Why? Often the lighting is too bright, and you can't look from the instruments to the sails or ahead into the darkness without losing your night vision for some moments. It's silly to have to tape paper or plastic over the instrument face at night to kill the glare, but sometimes it's either that or don't turn on the night lighting. Instruments with rheostat dimmers are a nice solution.

The other aspect of readability you don't get to

experience in advance is resistance to fogging. If the instrument has a sealed, double-layer lens, or is itself hermetically sealed, you should have no problem. Most units don't have these features, however.

INTEGRATED SYSTEMS

So much for basic wind and water indicators. But wait—we're not through with these functions; we have other uses for them. From them can be derived what are called computed functions, data that are the result of combining wind and water inputs and adding a dash of trigonometry. The most important of these computed functions are velocity made good (V_{mg}), true wind speed, and true wind direction.

Velocity made good tells you which course and speed is most efficient for getting to a point upwind or downwind. Will pinching the boat higher and trimming the headsail right to the spreader result in a net gain to windward, even if you're going noticeably slower on the speedometer, or should you crack the sheets a bit and foot off on a faster course that's not so close to the wind? Downwind, V_{mg} tells you which sailing angle is the most efficient to reach or run at.

From a safety standpoint, true wind direction and speed—the values you'd measure if you were sitting atop a buoy rather than on a moving boat—are more valuable still. If you're sailing directly downwind with the apparent wind speed indicator at 15 knots and the knotmeter at 8, you know the wind is blowing 23 knots true. That means you'd better have your stout headsails ready to hoist, and you should put a reef or two in the mainsail *before* turning upwind.

True wind direction means you can select the right sails more easily before heading off on a reach. Will you be able to set a spinnaker, or must you stay with the genoa or set a reacher and a staysail? This judgment is not easy to make before you've actually turned onto the new course, and if you've guessed wrong you must go through a needless sail change. True wind direction gives you an unadulterated guide to use in conjunction with a polar chart or table of wind angles matching the sails in your inventory to the appropriate conditions.

These functions are useful, but calculating them is usually far more time consuming than the simple upwind-downwind addition of wind speed. It wasn't until manufacturers put microprocessors into their wind and water instruments that such data were made available on demand. When it happened, though, it conjured a whole new way of judging sailing performance.

When you are sailing to windward, the apparent wind is greater than the true wind. Knowing the boat's speed, and the apparent wind's speed and direction, true wind can be computed, and true wind is the key to the other computed functions. It must be found first; then the others follow.

$$V_t = \sqrt{V_s^2 + V_a^2 - (2V_s \times V_a \times \cos D_a)}$$

$$D_t = D_a + \arcsin (V_s \times \sin D_a / V_t)$$

$$V_{mg} = V_s \times \cos D_t$$

where V_t = true wind velocity
V_s = boat velocity
V_a = apparent wind velocity
D_a = apparent wind direction
D_t = true wind direction
V_{mg} = boat velocity made good

The computer's memory supplies square root, arc sine, and cosine values, along with the simple math. To be really accurate, leeway must be accounted for in the true wind direction value, but that requires sensors not included in most computed function systems. V_t, D_t, and V_{mg} are the building blocks for a series of functions not used in the past. These include:

- Distance lost to weather on each tack
- Speed or distance gained to leeward
- Optimum theoretical boat speed based on polar speed curves of the boat's performance potential
- Difference between actual speed and theoretical potential
- Predicted apparent wind direction and speed for the next leg to be sailed
- Optimum tacking angles upwind and jibing angles downwind
- Distance made to the next waypoint
- V_{mc} or target speed. This is a different way of perceiving performance.

There are many more to choose from, and you can invent some of your own by programming a personal or portable computer that links to the system through a standard interface.

Computed functions come to you plain, fancy, and very fancy. The choices are varied, but in the most

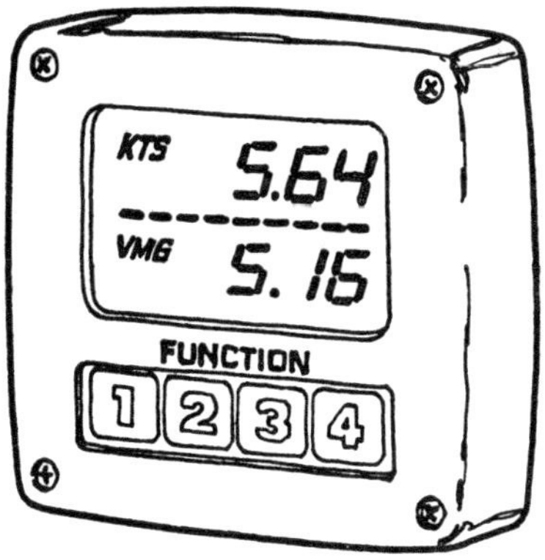

Digital multifunction displays allow readings from several sensors to be shown in one compact box. This box may simply display the data from the sensors, or it may use the data to calculate additional functions such as speed made good (V_{mg}), true wind speed, or true wind direction. In the latter case, the box is part of an integrated system, and the prices for such systems range from $800 up.

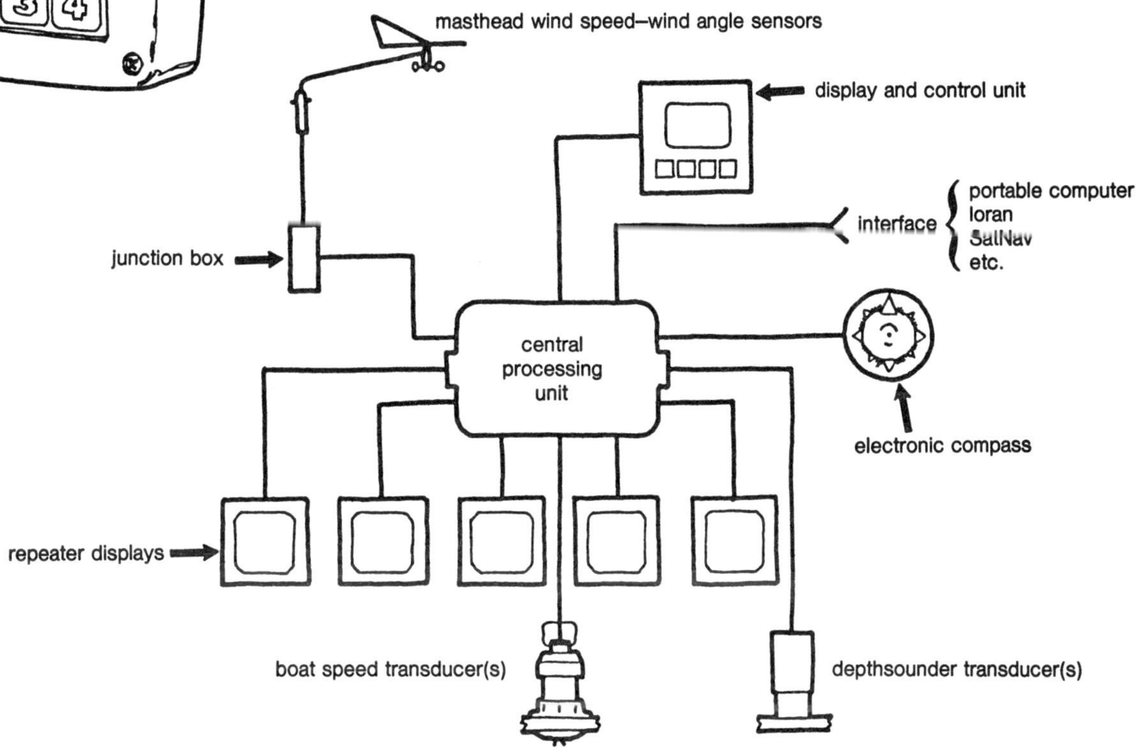

A central processor-based integrated system feeds inputs into the CPU from speed, wind, heading, depth, and heel sensors, and from outside units such as loran, SatNav, or a computer. The CPU then distributes the calculated data to the appropriate readouts in the nav station and cockpit. Another approach to system integration is to give each instrument enough independent computing power to use input from any other instrument. There being no CPU in this kind of system, the system as a whole may not go dead if a major part of it shuts down. Each unit can act independently or in concert. The CPU system has nevertheless performed well in the three major brands that use it.

general terms, instruments that produce these functions can be put into one of two groups: independent unit systems, and central processor unit (CPU) systems.

Independent units usually look and work like their basic one-function counterparts, but they have more plugs in the back and more buttons in the front. The plugs allow inputs from boat speed, wind speed, and wind angle transducers. The buttons let you call up, say, three to ten times as much information as you can get on the display of a comparable single-function model. A trend in displays is toward larger LCDs with graphics that supplement the alphanumerics. One will show you digitized pointers for both apparent or true wind direction, with wind speed in numbers below. It can show reference lines for tacking angles, and it flashes at you when you sail by the lee downwind. Such instruments take a little while to learn, and before buying one it would help to pick up a copy of the owner's manual and all the sales literature available. Then you'll have time to ponder its mysteries and decide whether you really want it.

The other group of instruments—CPU systems—includes all the cutting-edge electronics for sailing.

Alphanumeric

Race course mark coordinates, distances, and headings for Gold Cup, Olympic, or offshore courses.
Predicted apparent wind angle and speed for next leg.
Predicted true wind angle and speed for next leg.
Target boat speed for prevailing apparent wind—now and on next leg.
Recommended sail configuration for conditions and course.
Predicted mode to next mark—tack, reach, or jibe.
Optimum heading to next mark.
Optimum heading on next jibe or tack.
Optimum reaching course—accounting for leeway.
Distance and ETA to various marks and waypoints.

Graphic

True wind speed and direction fluctuations plotted over time.
Laylines to mark as wind shifts.
Tacking efficiency plot—for comparing with other watch.
V_{mg} plotted over time—also for comparison.
Current set and drift vectors.

Some computed data available in current software packages. These are only a few of the many functions that can be programmed or bought as ready-to-go software for computers such as the IBM PC, the PC Jr, the Apple IIe, the Epson HX 20, the Hewlett-Packard HP-41, the HP-75, the Compaq, the Data General One, and others. Consult the integrated system manufacturer for a list of compatible software and hardware.

This is the stuff carried aboard America's Cup 12 Meters, International Offshore Rule ocean racers, and transoceanic record-breaking multihulls. A CPU system is a bunch of transducers, sensors, and control units that feed a central processor. It then spews out information to system displays, a printer, or an outside computer. The computer can be programmed to display a whole range of special functions. (See the table for examples.) The CPU can also take inputs from a SatNav or loran receiver and display true position on the race course along with time, distance, and course to marks and laylines. It can compute current set and drift, or leeway. It has its own electronic compass, for which it memorizes compensations for deviation and variation, and it can display true or magnetic bearings and courses, along with their reciprocals.

It includes a host of calibrations and corrections for such things as heeling angle and upwash on the wind angle readouts. The CPU can also be programmed with a "trial horse." This is a table of performance targets in various wind and sea conditions based on your boat's design parameters and different sail combinations. You get a fairly realistic view of how well you're doing relative to the ultimate. It helps in tuning the boat—racer or cruiser—as well as during a race, when your competition is too far away to serve as a yardstick for judging the effects of your trim changes.

Some integrated systems do what CPU systems do without depending on a central processor. Each of the components functions independently, but also communicates with the others. The most advanced uses a single coaxial cable to carry system power as well as all the various instrument dialogues from component to component. Such systems are becoming popular with sailors who want system redundancies for safety's sake, or because they feel they can customize their installation more easily.

Installing, calibrating, and learning to use an integrated system may not be simple, but you should take the time to do as much of it as possible yourself so you can understand the system and use it to its potential. A wiring expert may save you trouble, however, by installing the components correctly the first time, with a minimum of resulting electrical interference. Most of the wiring connections in these systems are solderless, multiprong affairs that only go together one way. Following the manufacturer's instructions should not be difficult. It's isolating the sensitive electronics from engine, radio, and motor noise, putting the electronic compass in the best possible position, and compensating it to 1-degree accuracy that takes professional expertise or else great patience.

Calibrating the electronics in an integrated system is no different than the calibration of individual instruments. There's more to do, though, and the tuning takes more care, because inconsistencies accumulate and are compounded when they're allowed to enter the computations. Inconsistency is the key term, not inaccuracy. As long as you can get an instrument to measure speed or direction changes in linear fashion, as opposed to reading accurately at some values and with varying error at others, the amount of error is not important. The calibration circuit in the instrument will solve the constant discrepancy.

The most sophisticated systems offer calibration for twin through-hull speed transducers, for upwash, leeway, and heel error in the wind-angle transducer, and for compass variation and deviation. Twin speed transducers will probably not give the same readings, but by turning one in its tube until its reading matches the other, you can then do your buoy-to-buoy speed and distance calibration as for a single through-hull unit. Upwash is the error created by the headsails' effect on airflow over the wind vane at the masthead. The sails bend the air to give it an apparent reading farther aft than it really is. Masthead rigs suffer significantly more than do fractional rigs. Upwash calibration brings the wind forward to where it would be if the genoa weren't hoisted. You can make this calibration on standard wind angle meters, too, simply by comparing wind angles with and without the headsail hoisted, at various headings in a steady breeze.

Heel error is a predictable value, as is leeway, for a given boat type, and these factors are thus automatically dealt with through calibration. An electronic heel sensor—a potentiometer—is needed to complete the correction circuit.

Compass variation and deviation, automatically compensated for in top-of-the-line SatNav receivers, are not so simply adjusted in other instrument systems. In some, variation can be corrected at the factory if you tell the company where you will be sailing. Deviation can only be corrected on board, underway. It varies with heel, though, so simply motoring on different headings from a known point won't solve the problem to the degree needed on a sailboat with a full-race electronics setup. Compensating magnets within the compass are too close, and can make the heel error worse on some points. Instrument system manufacturers recommend larger magnets placed farther away from the compass, recalling the quadrantal spheres of iron placed on either side of the steering compass aboard the old steel square riggers.

At a recent electronics seminar, a potential buyer asked if, for his $7,000, his integrated system would come with installation and troubleshooting services included. The manufacturer answered him without blinking, "We have found that the best operating systems are the ones that are owner-installed." It was not a cop-out, but a statement of reality. Although these systems make sailing a boat with maximum efficiency easier, they also demand more effort on your part in installing, tuning, and understanding them.

You don't have to be a cost-is-no-object racer to get many of the advantages of performance indicators with computed functions. All you need to do is look at your requirements as dispassionately as possible, and then spend time at the dealers and boat shows in your area. It's not a simple purchase, so take your time. The choice, and the prices, can only get better.

BUYING TIPS

There is a wide range of choices in instrumentation, just as with any other piece of marine electronics. Prices vary, and so do accuracy, ruggedness, and dependability. At one time the cheapo stuff was easy to spot: It had the plastic indicator housing and through-hull unit. Now almost every manufacturer's electronics come in lightweight plastic cases, which in many instances are preferable for watertightness and ease of maintenance.

Can you tell anything about an instrument by picking it up and looking at it? Well, sometimes. Does the transducer paddlewheel or spinner rotate freely? How much breath is required to make it start to turn? Are the paddles well molded, with the magnets molded in and not just glued on? Is the transducer well "potted" in epoxy inside the through-hull tube? Are there connectors in the coax cable 5 to 10 feet from the through hull as well as at the indicator unit? Is the indicator itself well gasketed where it must face the elements? Is safety glass used for the dial, rather than something more breakable? Does the masthead wind speed and direction unit rotate at the faintest breath of air? How does it register on the indicator—instantly and wildly, or after a slight delay for damping? Does the indicator give full 360-degree wind coverage?

You can compare spec sheets, of course. Power draw, usually expressed in microamps. Accuracy, in percentage of true speed (such as ±2%) or in degrees of wind angle. Sensitivity—how light a breeze or how low a boat speed is sufficient for reliable readings—depends on bearing quality and lightweight moving parts. For masthead components, especially, this is important.

Dependability is best determined by looking at

instruments aboard other boats in your harbor and asking the owners for a brief operational history. Don't just ask, "How good is it?" The answer will usually be "Great!" as selective memory takes control and blocks out the bad experiences. Ask how many times the gear has gone down, and for what reason. Who fixed it: the dealer (usually meaning a minor problem) or the manufacturer (who deals with the big problems)? How long did it take to get back on line? Surveys of this type are more illuminating than all the dealer assurances you can get.

Where should you buy instruments and sailing computers? Is there any advantage in buying from a discounter? The question applies only to the simple, single-function equipment. This stuff is sold in large enough quantities to be sold below full markup at a profit. It is simple enough to install and test, and just about anyone can do the job. It thus makes sense to look for the lowest price on basic speedometers and wind instruments, assuming you're buying a well-known brand that's well supported by the manufacturer should something go wrong. For the more specialized gear, those electronics that require exceptional care in installation and calibrating, it's worthwhile to have the dealer around to help, though again, you should try to do at least the calibration part on your own as much as possible. Anyway, the integrated systems just don't get discounted. Pay the retail price, and expect retail service.

Appendix A: MAIL-ORDER DISCOUNT HOUSES

The following list will serve as a starting point for the mail-order bargain hunter. The larger, better known firms are included, as are companies of regional interest. Many warehousers also operate showrooms for the benefit of local customers. Check with those in your area.

Some of the mail-order houses, such as Goldberg's Marine and E&B Marine, are general merchandisers of all manner of boat gear, while others specialize in marine electronics. The only concrete advantage in ordering from members of the latter category, such as Skipper Marine Electronics or American Marine Electronics, is that you can get more or better information and advice over the telephone from them. With that exception, wise shopping through the mail-order houses consists of finding the best price on the item you want at the time you want it. Keep in mind that shipping and handling charges vary from one company to another and can alter the bottom line, although the effect is not often dramatic on high-ticket marine electronic items.

Although, to our knowledge, the companies listed here all provide satisfactory service, the list is not intended as an endorsement, nor is any company's exclusion from the list intended as a condemnation. The addresses and telephone numbers were current as of April 1985.

American Marine Electronics
2 Wilton Ave.
Norwalk, CT 06851
(800) 243-0264

BOAT/US
880 S. Pickett St.
Alexandria, VA 22304
(800) 336-0226

BOAT/US offers discount prices to members only. Cost of membership is $17 per year and provides access to information, documentation, and insurance services.

Coast Navigation
1934 Lincoln Dr.
Annapolis, MD 21401
(800) 638-0420

Crook & Crook
2804 Bird Ave.
Miami, FL 33133
(800) 432-2381

Defender Industries
P.O. Box 820
255 Main St.
New Rochelle, NY 10801
(914) 632-3001

E&B Marine
P.O. Box 747
980 Gladys Court
Edison, NJ 08818
(201) 287-3900

Findlay Imports
101-1429 Dominion St.
North Vancouver, British Columbia
CANADA V7J 1B3
(604) 985-8747

Fore and Aft Marine
12055 Seminole Blvd.
Largo, FL 33540
(813) 391-3559

Goldberg's Marine
202 Market St.
Philadelphia, PA 19106
(800) 262-8464

International Marine Supply
819 Pompton Ave.
Cedar Grove, NJ 07009
(201) 857-2646

Landfall Navigation
354 West Putnam Ave.
Greenwich, CT 06830
(203) 661-3176

Manhattan Marine & Electric Co.
116 Chambers St.
New York, NY 10007
(800) 221-1924

Marine Buyer's Co-op
207 Lincoln St.
Hingham, MA 02043
(800) 225-5244

Marine Center
1150 Fairview North
Seattle, WA 98109
(206) 682-1150

Martech
P.O. Box 14070
Ft. Lauderdale, FL 33302
(800) 327-2541

M&E Marine Supply Co.
P.O. Box 601
Camden, NJ 08101
(609) 858-1010

M.M.O.S.
3034 Malmo Dr.
Arlington Heights, IL 60005
(312) 952-8444
15219 Michigan Ave.
Dearborn, MI 48126
(313) 582-9480

National Marine Supply Corp.
Keystone Executive Plaza
12555 Biscayne Blvd., Suite 924
Miami, FL 33181
(800) 645-2565

Outer Banks Marine
Highway 70 East
P.O. Drawer 500
Beaufort, NC 28516
(800) 682-2225

The Ship's Store Distributors
501 E. Boston Post Rd.
Mamaroneck, NY 10543
(914) 698-0056

Skipper Marine Electronics
3170 Commercial Ave.
Northbrook, IL 60062
(800) 621-2378

Spa Creek, Inc.
616 Third St.
Annapolis, MD 21403
(301) 267-6565

Starboard Marine
3078 Lawson Blvd.
Oceanside, NY 11572
(800) 645-6541

Tom Taylor Co.
136 Adelaide St. East
Toronto, Ontario
CANADA
(416) 363-8586

Warehouse Marine
P.O. Box 70348
Seattle, WA 98107
(800) 426-8666

West Marine Products
P.O. Box 5189
2450 17th Ave.
Santa Cruz, CA 95063
(800) 538-0775

Appendix B: MANUFACTURERS & IMPORTERS

The following table lists the manufacturers or North American importers of the marine electronic instruments and systems covered in this book. The list is as complete and current as we could make it. If we have overlooked any companies, we have done so unintentionally.

	Depthsounders					Radio Direction Finders											
	Digital or meter	Flasher	Recording	VHF Radios	SSB Radios	RDF	ADF	VHF/ADF	Loran	SatNav	Autopilots	Radar	Anemometers	Wind Direction Indicators	Speedometers	Multiple Display Instruments	Integrated Systems
Alpha Marine Systems, 996 Hanson Ct., Milpitas, CA 95035. (408) 945-1155											✓						
Apelco, 676 Island Pond Rd., Manchester, NH 03103. (603) 668-1600		✓	✓	✓		✓		✓	✓								
Aqua Meter Instruments, 465 Eagle Rock Ave., Roseland, NJ 07068. (201) 228-3600	✓	✓	✓			✓									✓		
Arnav Systems, 4740 Ridge Dr. NE, Salem, OR 97303. (800) 547-0254	✓				✓				✓								
Atlantic Marine Div. IMI Industries, 40 Signal Rd., Stamford, CT 06902. (203) 357-8455	✓												✓	✓	✓	✓	✓
Baymar Marine Electronics, 1705 Enterprises Dr., Fairfield, CA 94533. (707) 429-9090			✓														
Bristol Electronics, 651 Orchard St., New Bedford, MA 02744. (617) 997-3181	✓			✓										✓	✓		
Brookes & Gatehouse, 154 E. Boston Post Rd., Mamaroneck NY 10543. (914) 698-9330	✓					✓				✓			✓	✓	✓	✓	✓
Canadian Marconi, 2442 Trenton, Montreal, Quebec, CANADA H3P 1Y9.				✓	✓												
Cetec Benmar, 3000 W. Warner Ave., Santa Ana, CA 92704. (714) 540-5120											✓						

	Depthsounders					Radio Direction Finders											
	Digital or meter	Flasher	Recording	VHF Radios	SSB Radios	RDF	ADF	VHF/ADF	Loran	SatNav	Autopilots	Radar	Anemometers	Wind Direction Indicators	Speedometers	Multiple Display Instruments	Integrated Systems
Cinkel Industries, 6220 Shawson, Mississauga, Ont. CANADA L5T 1J8. (416) 676-1234											✓						
Coast Navigation, 1934 Lincoln Dr., Annapolis, MD 21401. (800) 638-0420			✓			✓											
Coastal Navigator, 4312 Stone Way North, Seattle, WA 98103. (800) 426-7787	✓	✓	✓	✓		✓			✓				✓	✓	✓		
Coe Mfg., Div. of TIC, P.O. Box 463, Dana Point, CA 92629. (714) 830-4518											✓						
Combi Marine Div. IMI Industries, 40 Signal Rd., Stamford, CT 06902. (203) 357-8455				✓	✓	✓	✓		✓								
Communications Associates, 200 McKay Rd., Huntington Station, NY 11746. (516) 271-0800					✓												
Compass Electronics, P.O. Box 366, Forest Grove, OR 97116. (503) 357-2111	✓	✓															
CPT, 4336 NE 11th Ave., Ft. Lauderdale, FL 33334. (305) 564-1445											✓						
Cruising Designs, P.O. Box 151, Peabody, MA 01960. (617) 532-2712											✓						
Cybernet, 7 Powder Horn Dr., Warren, NJ 07060. (201) 560-0060				✓													
Danaplus APS, Ved Klaedebol, 2970 Hoersholm, Denmark 45-286-8289	✓												✓	✓	✓	✓	✓
Datamarine International, 53 Portside Dr., Pocasset, MA 02559. (617) 563-7151	✓		✓						✓				✓	✓	✓		

	Depthsounders					Radio Direction Finders											
	Digital or meter	Flasher	Recording	VHF Radios	SSB Radios	RDF	ADF	VHF/ADF	Loran	SatNav	Autopilots	Radar	Anemometers	Wind Direction Indicators	Speedometers	Multiple Display Instruments	Integrated Systems
D & D Electronics, 246 S. Main St., Albany OR 97321. (503) 928-5375		✓	✓														
Digital Marine Electronics, 30 Sudbury Rd., Acton, MA 01720. (617) 897-6600									✓								
R.L. Drake, 540 Richard St., Miamisburg, OH 45342. (513) 866-3621				✓	✓												
Electro Marine Systems, 96 Fox Hunt La., E. Amherst, NY 14051. (716) 688-6500	✓	✓											✓	✓	✓	✓	
Energy Control, P.O. Box 127, Bellevue, WA 98009. (206) 455-4666											✓						
Epsco Marine, 411 Providence Hwy., Westwood, MA 02090. (617) 329-1500									✓			✓					
First Mate Marine Autopilots, 41 Kindred St., Stuart, FL 33494. (305) 286-4480											✓						
Furuno, 271 Harbor Way, S. San Francisco, CA 94080. (415) 873-9393			✓	✓			✓		✓	✓		✓					
Jay Stuart Haft, P.O. Box 11210, Bradenton, FL 33507. (813) 753-6446	✓												✓	✓	✓		✓
Heath Co., Benton Harbor, MI 49022. (616) 982-3200	✓	✓	✓												✓		
Hull Electronics, 7563 Convoy Ct., San Diego, CA 92111. (619) 278-6140					✓												
ICOM America, 2112 116th Ave. NE., Bellevue, WA 98004. (206) 454-8155				✓	✓												

	Depthsounders					Radio Direction Finders											
	Digital or meter	Flasher	Recording	VHF Radios	SSB Radios	RDF	ADF	VHF/ADF	Loran	SatNav	Autopilots	Radar	Anemometers	Wind Direction Indicators	Speedometers	Multiple Display Instruments	Integrated Systems
II Morrow, P.O. Box 13549, 2777 19th St. SE, Salem, OR 97309. (503) 581-8101									✓		✓						
Impulse Mfg., 329 Railroad Ave., Pittsburg, CA 94565. (415) 439-2072	✓														✓		
Inmar Electronics & Sales, P.O. Box 671, Smithtown, NY 11787. (516) 543-0080	✓	✓															
Intech, 282 Brokaw Rd., Santa Clara, CA 95050. (408) 727-0500								✓		✓							
International Marine Instrument Div. IMI Industries, 40 Signal Rd., Stamford, CT 06902. (203) 357-8455	✓	✓							✓	✓	✓	✓	✓	✓	✓	✓	✓
ITT Decca Marine, P.O. Box G, Palm Coast, FL 32037. (904) 445-2400	✓	✓	✓						✓								
Kenyon, P.O. Box 308, Guilford, CT 06437. (203) 453-4374	✓												✓	✓	✓	✓	✓
King Marine Radio, 5320 140th Ave. North, Clearwater, FL 33520. (813) 530-3411	✓	✓	✓	✓	✓				✓						✓		
Lowrance Electronics, 12000 E. Skelly Dr., Tulsa, OK 74128. (918) 437-6881	✓	✓	✓														
Magnavox Advance Products & Systems, 2829 Maricopa St., Torrance, CA 90503. (213) 618-1200										✓							
Mars Electronics, Marine Sys. Div., 801 Carpenters Crossing, Folcroft, PA 19032. (215) 534-4200												✓					
Metal Marine Pilot, 2119 W. Mildred St., Tacoma, WA 98466. (206) 564-5902											✓						

	Depthsounders					Radio Direction Finders											
	Digital or meter	Flasher	Recording	VHF Radios	SSB Radios	RDF	ADF	VHF/ADF	Loran	SatNav	Autopilots	Radar	Anemometers	Wind Direction Indicators	Speedometers	Multiple Display Instruments	Integrated Systems
Micrologic, 20801 Dearborn St., Chatsworth, CA 91311. (818) 998-1216									✓								
Mieco Div. Polarad Electronics Corp., 109 Beaver Ct., Cockeysville, MD 21030. (301) 667-4660									✓								
Monitor Devices, 850 Towbin Ave., Lakewood, NJ 08701. (201) 370-0700	✓										✓						
Motorola Marine, 1301 E. Algonquin Rd., Schaumburg, IL 60196. (312) 576-3632					✓												
Nautical Electronics Co. (NELCO), 7095 Milford Ind. Rd., Baltimore, MD 21208. (301) 484-3284									✓								
Navidyne, 11824 Fishing Pt. Dr., Newport News, VA 23606. (804) 874-4488									✓	✓							
Navigation Instruments, 1934 Lincoln Dr., Annapolis, MD 21401. (301) 268-3120						✓											
Navstar, 756 Lakefield Rd., Bldg. J, Westlake Village, CA 91361. (805) 495-8070										✓							
Newmar, P.O. Box 1306, Newport Beach, CA 92663. (714) 751-0488						✓											
Ockam Instruments, 122 Cascade Blvd., Milford, CT 06460. (203) 877-7453	✓												✓	✓	✓		✓
Paragon Electronics, P.O. Box 1456, Bellevue, WA 98009. (206) 454-0846	✓																
Pearce-Simpson, Gladding Corp., P.O. Box 520800 GMF, Miami, FL 33152. (305) 693-9113	✓	✓		✓	✓	✓											

	Depthsounders					Radio Direction Finders											
	Digital or meter	Flasher	Recording	VHF Radios	SSB Radios	RDF	ADF	VHF/ADF	Loran	SatNav	Autopilots	Radar	Anemometers	Wind Direction Indicators	Speedometers	Multiple Display Instruments	Integrated Systems
C. Plath, North American Div. of Litton Sys., 222 Severn Ave., Annapolis, MD 21403. 800-638-0428						✓	✓			✓	✓	✓					
Racal Marine, 14778 NE 95th St., Redmond, WA 98052. (206) 882-0731					✓					✓	✓	✓					
Radar Devices, 2955 Merced St., San Leandro, CA 94577. (415) 483-1953												✓					
Radio Holland U.S.A., 6033 South Loop East, Houston, TX 77033. (713) 649-1048				✓	✓												
Rauff & Sorensen, 495 Post Rd. East, Westport, CT 06880. (203) 222-9600	✓			✓	✓					✓							
Ray Jefferson, Main & Cotton Sts., Philadelphia, PA 19127. (215) 487-2800	✓	✓	✓	✓		✓	✓				✓		✓	✓	✓		
Raytheon Marine, 676 Island Pond Rd., Manchester, NH 03103. (603) 668-1600	✓	✓	✓	✓	✓	✓			✓	✓		✓					
RB II, 455 Cayuga Rd., Buffalo, NY 14225. (716) 632-7474	✓												✓	✓	✓	✓	
Regency Electronics, 7707 Records St., Indianapolis, IN 46226. (317) 545-4281	✓			✓		✓											
RF Communications Group, Harris Corp., 1680 University Ave., Rochester, NY 14610. (716) 244-5830				✓	✓												
Robertson Div., Kongsberg N. America, 400 Oser Ave., Hauppauge, NY 11788. (800) 645-3738											✓						
Rochester Technologies, 9474 Chesapeake Dr., San Diego, CA 92123. (619) 279-5766									✓							✓	✓

	Depthsounders					Radio Direction Finders											
	Digital or meter	Flasher	Recording	VHF Radios	SSB Radios	RDF	ADF	VHF/ADF	Loran	SatNav	Autopilots	Radar	Anemometers	Wind Direction Indicators	Speedometers	Multiple Display Instruments	Integrated Systems
Ross Labs, 3138 Fairview Ave. E., Seattle, WA 98102. (206) 324-3950	✓	✓	✓														
SBA Sideband Assoc., 115 Phelam Ave., Ste. 7, San Jose, CA 95112. (408) 275-1150					✓												
Scanmar Marine Prod., 298 Harbor Dr., Sausalito, CA 94965. (415) 332-3233											✓	✓					
Seatron Marine Electronics, 4312 Main St., Philadelphia, PA 19127. (215) 487-2800			✓		✓												
SGC, 13737 SE 26th St., Bellevue, WA 98005. (206) 746-6310					✓												
Signet Scientific, 3401 Aerojet Ave., El Monte, CA 91731. (213) 571-2770	✓												✓	✓	✓	✓	✓
Simrad, 2208 NW Market St., Seattle, WA 98107. (800) 426-5565	✓		✓			✓	✓		✓								
Si-Tex Marine Electronics, P.O. Box 6700, Clearwater, FL 33518. (813) 535-4681	✓	✓	✓	✓			✓		✓	✓		✓	✓	✓	✓		
Southern Marine Research, 1401 NW 89th Ct., Miami, FL 33172. (305) 591-9433	✓	✓	✓	✓	✓												
SR Instruments, 600 Young St., Tonawanda, NY 14150. (716) 693-5977	✓												✓	✓	✓	✓	
SRD Labs, 381 McGlincey Ln., Campbell, CA 95008. (408) 371-2666									✓								
Standard Communications, P.O. Box 92151, Los Angeles, CA 90009. (213) 532-5300	✓			✓											✓		

	Depthsounders					Radio Direction Finders											
	Digital or meter	Flasher	Recording	VHF Radios	SSB Radios	RDF	ADF	VHF/ADF	Loran	SatNav	Autopilots	Radar	Anemometers	Wind Direction Indicators	Speedometers	Multiple Display Instruments	Integrated Systems
Stephens Engineering Assoc., 7030 220th St. SW, Mountlake Terrace, WA 98043. (206) 771-2182					✓												
Techsonic Industries, One Humminbird La., Eufaula, AL 36027. (205) 687-6615		✓	✓														
Telco Products, 25 Lumber Rd., Roslyn, NY 11576. (516) 484-6688				✓													
Telcor Instruments, 17785 Sky Park Circle, Irvine, CA 92714. (714) 250-1016	✓												✓	✓	✓		
Tele Comm Communications, P.O. Box 3232, Margate, NJ 08402. (609) 822-8588					✓												
Texas Instruments, P.O. Box 405, MS 3438, 2501 S. Hwy. 121, Lewisville, TX 75067. (214) 462-5220				✓	✓				✓	✓							
Tiller Master, 774 West 17th St., Costa Mesa, CA 92627. (714) 675-1668											✓						
Tracor Instruments Div., 6500 Tracor Lane, Bldg. 27, Austin, TX 78221. (512) 929-2051										✓							
Trans World Communications, 240 Pauma Pl., Escondido, CA 92025. (619) 747-1079					✓												
Trimble Navigation, 1077 Independence Ave., Mountain View, CA 94043. (415) 962-9893									✓								
Uniden Corp., 15161 Triton Lane, Huntington Beach, CA 92649. (714) 898-0558		✓		✓													
Unipas, 1142 Cranston St., Cranston, RI 02920. (401) 943-6800											✓						

	Depthsounders					Radio Direction Finders											
	Digital or meter	Flasher	Recording	VHF Radios	RDF	SSB Radios	ADF	VHF/ADF	Loran	SatNav	Autopilots	Radar	Anemometers	Wind Direction Indicators	Speedometers	Multiple Display Instruments	Integrated Systems
VDO Instruments, 980 Brooke Rd., Winchester, VA 22601. (703) 665-0100	✓												✓	✓	✓		
Viking Instruments, 400 Lake St., Kingston, MA 02364. (617) 426-5661												✓					
Wesmar, Box C3001, Bothell, WA 98041. (206) 481-2296			✓								✓	✓					
WH Autopilots, 655 NE Northlake Pl., Seattle, WA 98105. (206) 633-1830											✓						
Yaesu Electronics Corp., P.O. Box 49, Paramount, CA 90723. (213) 633-4007				✓	✓												
Yazaki Corp., Route 4, Airport Rd., Princeton, MN 55371. (612) 389-2303	✓		✓										✓	✓	✓		